UMTS

UMTS
Mobile Communications for the Future

Edited by

Flavio Muratore

CSELT, Telecom Italia Group, Italy

JOHN WILEY & SONS, LTD
Chichester • New York • Weinheim • Brisbane • Singapore • Toronto

First published under the title
LE COMUNICAZIONI MOBILI DEL FUTURO – UMTS: il nuovo sistema del 2001
© 2000 - CSELT - Centro Studi e Laboratori Telecomunicazioni S.p.A., via G. Reiss Romoli, 274 – 10148
Torino

Copyright © 2001 by John Wiley & Sons, Ltd,
           Baffins Lane, Chichester,
           West Sussex PO19 1UD, England

           National: 01243 779777
           International: (+44) 1243 779777

e-mail (for orders and customer service enquiries): cs-books@wiley.co.uk

Visit our Home Page on http://www.wiley.co.uk/ or http://www.wiley.com/

Reprinted February 2001, March 2001

Other Wiley Editorial Offices

John Wiley & Sons, Inc., 605 Third Avenue,
New York, NY 10158-0012, USA

WILEY-VCH Verlag GmbH,
Pappelallee 3, D-69469 Weinheim, Germany

Jacaranda Wiley Ltd, 33 Park Road, Milton,
Queensland 4064, Australia

John Wiley & Sons (Canada) Ltd, 22 Worcester Road,
Rexdale, Ontario M9W 1L1, Canada

John Wiley & Son (Asia) Pte Ltd, 2 Clementi Loop #02-01,
Jin Xing Distripark, Singapore 129809

Library of Congress Cataloging-in-Publication Data
Comunicazioni mobili del futuro–UMTS, English
   UMTS : mobile communications for the future / edited by Flavio Muratore.
     p.cm.
   Includes bibliographical references and index.
   ISBN 0–471–49829–7
     1. Mobile communication system EMI. Muratore, Flavio II. Title.
   TK6570.M6 C65413 2000
   621.3845–dc21   00–043914

British Library Cataloguing in Publication Data
A catalogue record for this book is available from the British Library

ISBN 0 471 49829 7

Typeset in Sabon by Deerpark Publishing Services Ltd., Shannon
Printed and bound by Biddles Ltd, Guildford and King's Lynn, UK.

This book is printed on acid-free paper responsibly manufactured from sustainable forestry, for which at
least two trees are planted for each one used for paper production.

D
621.3845
Com

Contents

Preface

Commercial start-up of the third-generation mobile system is scheduled for the year 2002. The name given to this system, at least in the European context, is the universal mobile telecommunication system (UMTS). Of the original idea conceived in the early 1980s, what remains today are the ambitious service features that the system must provide to the user: the ability to communicate in movement, anytime and anywhere, through an enormous variety of applications and universally usable terminals. These expectations are attracting increasing attention from the mass media, and are seen by the public at large as the natural evolution of a process which in a few short years has enabled the cellular telephone to enjoy a success that few would have thought possible.

The mobile systems that we have now come to take for granted have done much to change how we live and communicate. Together with the potential offered by the Internet, they have even changed some of our ways of thinking, at levels that are far deeper than might seem at first sight. How we work, use information, represent concepts and exchange messages have all changed. To an ever-increasing extent, the new media bring together voice, images and data, or even make these different communication modes interchangeable. This is possible because of the common digital representation shared by information content, and the synthesis and coding techniques associated with it.

Thus, the UMTS system springs from convergence between the worlds of telecommunications and information technology. The new mobile system could well prove to be an ambitious synthesis of the evolution of these two worlds, especially at the level of services.

Aside from the shared expectations, however, preparing the way for the UMTS system has been a far from straightforward process, and many of the system's basic aspects are still open to different interpretations and solutions. At the moment, for example, specifications are addressing at least three different radio interface modes, two of which have been adopted at the European level.

These different ways of responding to a shared vision of UMTS reflect the variety of interests at stake, and the unequal rates and stages of evolution in the countries involved. The different stances that have been taken up regarding the system's implementation are confirmed, however involuntarily, by the first letter in its acronym, which stands, not for unique, but for universal. And this latter characteristic is most likely to be achieved by making different techniques compatible at the service level, rather than by developing a single solution for all continental regions.

Be this as it may, the system's complexity and the enormous economic interests hinging on it have led to the consolidation of certain technological and systems-related aspects, where a common vision now prevails. The major innovations that have been achieved range from the service creation approach to the associated features' independence of the network layers, and to the flexibility of the transport functions, which can cover a broad range of application requirements. In addition, the UMTS system's evolution is seen as a continuation of existing systems and services. The new system, in fact, grows from a mobile market that is now firmly consolidated, at least with regard to voice services. GSM operators, who have deployed (and continue to invest) massive financial resources and know-how in the complexities of specifying the system, are aiming at a relatively graceful transition (a sort of soft handover, as it were) from today's system to UMTS. Indeed, the UMTS specifications acknowledge this need for gradual migration by calling for multi-mode terminals and the adoption of network architectures that are largely derived from GSM solutions.

Today, the standards-writing groups in Europe, Japan and to some extent in the United States are collaborating in defining a system which, if not unique, can truly be termed universal. This degree of convergence is by no means accidental, and has largely been achieved through the determination shown by TIM in its strategic contributions at the international level.

This book deals chiefly with the technical and service solutions that have been adopted in this context. Though the topics covered are highly specialised by nature, every effort has been made to ensure that the basic concepts are accessible to a wide readership, as the book is addressed to decision makers in related industries in addition to those working in the specific technical sectors concerned.

There can be no doubt that the book is one of the first to be published on the topic. With specifications still in a state of flux, any such effort to organise the many issues involved and put them in context is of enormous value, as it provides a consistent view of the entire system and the services it is expected to support.

The preview of the UMTS system's content, technical scenarios and services that the book offers has been made possible by TIM's early commitment to drawing up specifications for UMTS, and the importance which the operator has from the outset assigned to meeting this new challenge. A significant part of this commitment was channelled through CSELT, which was directly involved in developing specifications and in assessing and testing candidate solutions. CSELT was thus able to consolidate its mastery of the mobile systems of the near future, building up a broad-based understanding of these systems and operative skills of great value. This is no mean achievement, if we think of the vital impact that this know-how can have on our country's growth prospects.

Cesare Mossotto
Torino
January 18, 2000

About the authors

The authors are CSELT researchers who have been active for a number of years in specifying terrestrial and satellite mobile radio systems and optimising their performance. The editor, Flavio Muratore, received his degree in electronic engineering from the Politecnico di Torino, and has over ten years experience at CSELT in the field of mobile radio systems, occupying positions of responsibility in standards-writing organisations and in international co-operative projects.

List of Contributors

The following CSELT authors contributed to this publication:

Flavio Muratore (editor)
Sergio Barberis
Valerio Bernasconi
Ermanno Berruto
Loris Bollea
Enrico Buracchini
Andrea Calvi
Giorgio Castelli
Antonio Cavallaro
Giovanni Colombo
Daniele Franceschini
Andrea Magliano
Nicola Pio Magnani
Bruno Melis
Antonella Napolitano
Giovanni Romano
Enrico Scarrone

1

Introduction

Flavio Muratore

There can be no doubt that mobile telephony and data transmission on the Internet were the two outstanding successes in telecommunications during the closing years of the century, and there is every sign that these successes will be no more than the starting point for those of the new millennium.

For a number of years, in fact, development work has focused on new 'third generation' systems, or in other words, systems with the enhanced capabilities needed to make user mobility compatible with the growing demand for multimedia communication.

Given the success of mobile telephony, the world's major players in telecommunications and the information society are working to specify these new third-generation mobile systems. In Europe, specifications have been drawn up for UMTS (*Universal Mobile Telecommunications System*), which will be a significant innovation over today's systems because of its high operating flexibility, its ability to provide a wide range of applications and, more generally, to extend the services now provided to fixed network users to mobile customers. What, however, are the driving forces behind this move to develop new mobile communication systems? What exactly are these systems, and how are they organised? What kind of services can they give us? How will today's terminals change?

This book will attempt to provide an answer to these and other questions.

Mobile radio systems have now reached levels of usage which few people would even have dared imagine just a few years ago.

Around the world, some 400 million people use these systems, with penetration levels that already exceed 50 percent of the population in certain countries.

At the same time, these systems' geographical radio coverage has far outstripped the most optimistic expectations, and some of the systems are present in a large number of countries. GSM, for instance, now extends well beyond the borders of Western Europe, the area for which it was originally conceived.

The most recent forecasts indicate that, by the end of the year 2003, there will be over one billion mobile terminals in operation around the world, which also means that they will exceed the number of fixed telephone lines foreseen for that date (as indeed is already the case in certain areas such as Italy).

On the Internet front, around 18 million new users log on every month, while data traffic doubles every six months or so. At this rate, it is clear that the Internet is becoming the most important channel for collecting and distributing information throughout the world.

A new era of multimedia communication, whereby voice, text and video can be combined in the same call, is rapidly becoming a reality in the world of mobile communications, where growth prospects are nothing if not excellent.

The new sector of multimedia mobile communications will make it possible to combine ongoing work on mobile telephony and the Internet in a single, concerted effort which will give the growth potential of the two areas – already brilliant when taken separately – a further boost.

The revolution that has taken place in the world of telecommunications over the last few years has not only changed our habits and lifestyles, but has also changed the outlook for developing countries, who quite rightly see access to telecommunications as one of the keys to economic and social success.

The time is now ripe for a further move forward, both because this is what people want and expect, and because the state of the art now makes such a move possible. Increasing numbers of people want access to information on the move, and want this information to cover a wider and more variegated range than can currently be provided.

For example, market surveys indicate that the demand for visual information continues to grow. At the moment, images can be acquired and transferred, stored in memory and processed, using commercial devices such as video camcorders, personal computers and cameras. These new tools brought to us by digital technology can be used to send 'electronic postcards' in real time, view potential purchases located anywhere in the world, share moments in our lives with distant friends and relatives, or to help people who are hurt, lost or are otherwise in distress. We will also be able to look up flight schedules and timetables for other forms of transportation, check our bank accounts and make remote payments with procedures that are simpler and more straightforward than those that are beginning to be available to us today.

If we look at what is happening around us now, it is clear that the new age of multimedia mobile communications has already begun.

On the Internet, a large number of multimedia applications are already available today. For instance, we have tele-working applications that make it possible to manage voice and text simultaneously, or to share documents and video clips that can be updated or edited by several users at the same time. There are applications that permit simultaneous communication between multiple users, e-commerce or stock trading. The latter kinds of transaction, in fact, are gradually ousting more traditional ways of doing business. Other examples of interactive services include latest-generation video games, where several players in different places can interact in a three-dimensional virtual environment, or applications that make it possible to choose films, radio channels or TV programs in real time.

Alongside these developments on the Internet, many companies have set up their own internal networks – or intranets – to manage the information and documents they produce using the same methods and applications as are used on the Internet.

In the area of mobile radio systems, new services based on limited-capacity Internet access are gradually finding their way onto the market.

Along these lines, the GSM networks are now being upgraded in order to make it easier to introduce data and multimedia services. This process is proceeding on a number of levels.

On the radio level, though wider frequency bands are not available for this purpose, attempts are being made to introduce functions which will provide greater flexibility in assigning physical resources to users, thus permitting communication to take place at rates above the traditional 9.6 kbit/s when required. Hence the *High Speed Circuit Switched Data* (HSCSD) solution, which gives GSM users bits rates up to eight times higher than those used today.

By the end of the year 2000, the *General Packet Radio Service* (GPRS) will, as the name implies, provide packet switched data services with transmission capacities up to 171 kbit/s. The *packet switching* mode, as opposed to the *circuit switching* mode in current use, occupies the GSM network's transmission resources only when there is information to be transmitted: the user pays only for the actual amount of information sent, rather than for the entire time that the communication remains active.

All of these efforts are intended to permit permanent connection to the Internet or to corporate intranets at affordable cost. In addition, special application platforms have been developed in order to provide mobile terminals with interfaces open to this kind of application. The *Wireless Application Protocol* (WAP), for instance, makes it possible to adapt information received from the Internet environment so that it can also be used on mobile radio terminals, overcoming the transmission capacity and graphic representation limits that respectively affect the radio link and the terminal. The GSM *Short Message Service* (SMS), which has been active for some years now, is being updated and, thanks to more advanced terminals, can already associate graphics with the messages it transmits. In the near future, there is no reason that we will not be able to attach voice messages or video clips, as can now be done with e-mail over the Internet.

Nevertheless, it is clear that there are insurmountable physical and functional limits to second-generation systems.

The real breakthrough will come only with new systems designed specifically for multimedia applications. The challenge which has been occupying us for some time is thus that of developing the so-called 'third generation' mobile radio systems.

The next growth wave will come with the enormous variety of services that a universal system will be able to offer. The universal availability of services is an essential prerequisite for making full use of the world telecommunication market's potential. These are some of the reasons which led to world-wide agreement that the new-generation system must be truly innovative with respect to the mobile radio systems now in service. The new system, for instance, must be able to provide services and performance features which come increasingly close to those made available by fixed systems, in terms of both quality of service and transmission rates. It is thus clear that fixed and mobile services must move towards convergence, at least as regards how they are used.

Multimedia services are extremely heterogeneous in terms of the demands they make on the telecommunications network. For example, applications with a high degree of interactivity (such as video conferencing and voice services) require that end-to-end delay be as limited, and as constant, as possible, whereas other applications such as e-mail do not make stringent demands on network delay. Many applications spring from the world of computer networks (e.g. file transfer), others from telecommunications (e.g. video broadcasting), and yet others from the Internet (e.g. e-business). Consequently, third-generation systems must be able to manage and provide users with a broad range of applications with widely differing technical requirements (delay, errors, privacy).

In view of the massive investments that operators have made in today's second-generation systems, however, their third-generation successors must also ensure a high level of backward compatibility.

As regards the international standardisation issues, the ITU (*International Telecommunication Union*) has been active for a number of years in standardising FPLMTS (*Future Public Land Mobile Telecommunication Systems*), which in 1997 was renamed IMT-2000 (*International Mobile Telecommunications-2000*). As early as the autumn of 1996, ITU-R TG 8/1 had specified the minimum requirements which a radio technology must satisfy in order to be considered as a candidate for IMT-2000. CSELT, accredited by the Ministry of Communications as the Italian delegation, contributed actively to the discussion, promoting co-ordination between the European delegations. As a result, CSELT was instrumental in seeing that the requirements approved were such as to ensure that IMT-2000 can provide features which are truly innovative by comparison with second-generation systems. Again in the autumn of 1996, agreement was reached concerning the road map and timeframes for specifying the IMT-2000 radio interface. ITU is thus the focal point for the work being carried out by the *regional* standards-writing organisations (such as the *European Telecommunications Standard Institute* – ETSI) in specifying third-generation mobile radio systems. Following ITU's lead, in fact, the various standardisation bodies have proposed solutions complying with the minimum requirements established for the IMT-2000 systems. In the last few years, a major effort has been made to ensure that the UMTS system now being specified in Europe by ETSI is also in line with the work done at ITU. To guarantee that a standard capable of achieving world-wide acceptance can be developed, close contacts have been maintained between ETSI and the other standardisation groups, such as ARIB (*Association of Radio Industries and Business*) in Japan, TTA (*Telecommunications Technology Association*) in Korea and ANSI in the United States.

These 'collaborations' have led to the creation of the so-called *Partnership Projects* (PPs), where experts from the various *regional* standards-writing groups participate in order to converge towards a technical solution with the highest possible degree of harmonisation.

Though ITU's original intent was to specify a single third-generation system, the presence of many players on the scene, who are motivated by highly dissimilar interests, has made it impossible to achieve this goal in full. It is now certain that IMT-2000 will consist of a family of systems (including the European UMTS solution) with a high level of compatibility and which will in any case be able to ensure world-wide roaming through the use of multi-mode terminals that can operate with European and American network standards.

As was mentioned above for third-generation systems in general, UMTS is a real innovation by comparison with present-day systems. The system is designed to be highly flexible, to provide a wide range of applications in a multitude of environments and, more generally, to extend most of the services now provided to fixed network customers to mobile users.

UMTS will be deployed in several successive stages, gradually increasing the capabilities it provides while guaranteeing backward compatibility with previous versions. The minimum objective is to provide services with bit rates up to at least 2 Mbit/s for low-mobility users, and up to 144 kbit/s for high-mobility users (the idea is to be able to operate with the mobile terminal moving at speeds up to 500 km/h). There is no intention, however, of closing the door on higher bit rate services, and interworking can be expected with W-LAN systems capable of providing services at bit rates in the order of several dozen Mbit/s through special terminals and access methods.

Another major concern is compatibility between the third-generation UMTS and the second-generation GSM system.

In view of GSM's enormous success, and given that UMTS is planned to enter commercial service in Europe starting in 2002, we are likely to see a gradual migration from GSM to UMTS. It is also likely that UMTS coverage will at the outset be provided only in high traffic areas or those with particular customer needs: user mobility over more extensive areas will be ensured – though with reduced services – by means of GSM. In such a scenario, it is thus necessary that UMTS terminals also be able to operate as GSM terminals, or in other words that they be dual mode units. Finally, it should be borne in mind that though UMTS sets out to provide global coverage (and thus to include the oceans and deserts, where using the satellite segment is indispensable), most current specification work concentrates only on the system's ground-based segment, called UTRAN (*UMTS Terrestrial Radio Access Network*).

A factor of fundamental importance for the future growth of third-generation mobile radio systems is the band made available to them. The band assigned to the IMT-2000 system, and thus to UMTS in Europe, has been established on the basis of a number of decisions taken at the *World Radio Conferences*, or WRCs.

The assigned bands are those between 1885 and 2025 MHz and 2110–2200 MHz.

The resolution that assigns bands to IMT-2000 does not establish the duplexing method, i.e. how the received signal is separated from the transmitted signal, which depends on the radio solution chosen for the system. It goes without saying, however, that the method will draw on solutions that have already been adopted in existing systems, such as the FDD (*Frequency Division Duplexing*) approach which achieves bi-directional transmission on two separate and symmetrical bands for the two links (as in GSM, for example), and TDD (*Time Division Duplexing*), where bi-directionality is accomplished through time division on the same carrier (as for DECT – *Digital European Cordless Telephone*).

The national regulatory bodies will be responsible for taking steps to ensure that the bands in question are in fact available when they are used by other systems.

On the basis of current market projections, in any case, it can reasonably be assumed that these bands will not be sufficient to provide all of the services that will be required: forecasts indicate that further band assignments will be needed in the years 2005 and 2010.

The service requirements outlined in this brief introduction have a major impact on the technological choices (for both radio and the network) that must now be made for UMTS. The book will provide an exhaustive description of these choices in nine chapters, of which this introduction is the first. Chapter 2 then illustrates the innovative aspects of the system, while Chapters 3, 4 and 5 explain the operating principles of the physical layer, the access network and the core network, respectively. Chapter 6 analyses the opportunities for the satellite segment with UMTS, and Chapter 7 assesses the evolutionary pathways for services and terminals in the multimedia context. Chapter 8 discusses the issues involved in carrying out field trials for new radio systems, and the final chapter – Chapter 9 – concludes the book with a look at the next steps in the system's evolution.

To help the reader come to grips with a subject that can often be highly specialised, the more technically oriented chapters are organised on two levels. The first introductory level illustrates the topic, explains the reasons underlying certain decisions, and describes the structure of the various functional layers and the links between them. The second level provides a closer look at aspects which in the authors' view deserve further attention. These sections are identified by a magnifying glass logo.

2

The New Service Requirements and the Factors Behind Innovation

Ermanno Berruto and *Giovanni Colombo*

2.1 The reasons for innovation

The factors which in recent years have led to the need to specify a mobile communications system showing innovative features with respect to currently available solutions include:

- New service requirements which cannot be covered by second-generation systems such as GSM.

- The availability of new radio bands addressed by existing operators and potential new competitors.

Evolution towards UMTS can best be understood by starting from an analysis of the requirements set by the sector's major players, or in other words, by potential users, network operators, service providers, manufacturers and regulatory bodies.

As part of the general demand for connectivity and seamless services, the market for mobile communications applications will be increasingly affected by the catalyst provided by Internet-based data services. The mobile multimedia services market is expected to grow sharply around 2003–2005. This upswing will also be facilitated by the emergence of new terminals, or smart phones, which will combine the typical capabilities of cellular systems with *Personal Digital Assistant* type functions and with multimedia service profiles.

To an ever increasing extent, the *user* demands:

- Seamless Internet–Intranet access.

- A wide range of available services, including multimedia types.

- Compact, lightweight and affordable terminals.

- Simple terminal operation.

- Open, understandable pricing structures for the whole spectrum of available services.

Obviously, the user's demands are expressed in terms of service – and service quality – and are independent of the particular technology adopted, providing that it does not affect the level of service.

Over the last few years, *network operators* and *service providers* (in Italy, the two roles are almost invariably filled by the same player) have concentrated their investments on extending the services provided by GSM, improving its performance, and keeping offerings up to date. The process of radio coverage is now moving towards stability: most European operators focus their efforts on updating services and applications. Operators with a good foothold on the market see the evolutionary pathway towards UMTS as a combination of innovative breakthroughs in radio access and in the fixed, or core, network which will be able to permit graceful migration from GSM to the new mobile communication system. These operators are thus aiming for a system that will be able to respond effectively to changing demand through:

- A radio access which minimises quality-constrained network development costs.

- Joint specification of certain interfaces (and in particular the interface between the access network and the core network), which will make it possible to maintain a multi-vendor environment (establishing accepted specifications for interfaces makes it possible for the operator to integrate equipment from different manufacturers in the network).

- Simple and effective systems for developing services.

- The ability to provide services to other operators' customers who are roaming on the network.

Manufacturers see the move to UMTS as a major industrial opportunity. The manufacturing industries who have been able to conquer a significant share of the GSM market intend to put their entire support behind this evolution, as the natural continuation of GSM, in terms of equipment and architectural solutions. Manufacturers who do not have a solid position on the GSM market, on the other hand, are aiming at a UMTS solution which will introduce real differences with respect to second-generation systems.

The goal of the *regulatory bodies* can be summed up in the need to create competitive market conditions through a multi-operator and multi-service system, and to manage the available radio spectrum fairly, ensuring equal operating conditions for all of the players on the scene. To guarantee such conditions, the regulator's efforts are being concentrated to an ever-increasing extent on the radio bands, the services, and the conditions for integration between the various operators, independently of the technical solutions that come to the fore.

2.2 The requirements for the UMTS system

The success that mobile communications have enjoyed in recent years is based almost entirely on voice services. Indeed, a number of forecasts maintain that voice services will continue to have a predominant role until sometime around 2005. It must be emphasised, however, that data applications (and those associated with the Internet world in particular) will show extremely high growth rates, similar to those predicted for equivalent fixed network data services.

Ever since the early '90s, mobile services have grown at an entirely unforeseen pace. In this general rise, the case of Italy is emblematic. Figure 2.1 shows a comparison of the growth of mobile communications in Italy with that of the other European countries: in 1998, the percentage of Europe's population which used mobile phones varied

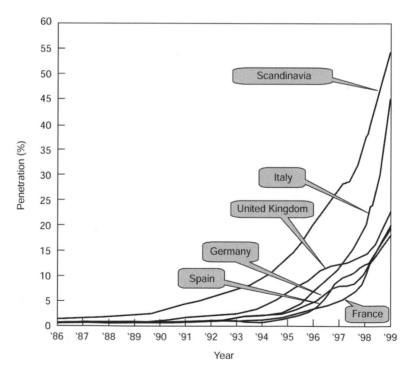

Figure 2.1   The growth of mobile communications in Europe.

between 20 and 40 percent, and exceeded 50 percent in certain countries.

In Italy, there can be no doubt that such impressive growth was bolstered by the decision to adopt a system such as TACS (*Total Access Communication System*) which from the beginning of the '90s onwards was able to offer all of the service 'portability' features that the potential mobile user expected in terms of terminal size and weight and battery life. The success of the TACS analog system demonstrates the extent to which service evolution depends on the user's direct perceptions (or needs), rather than on the specific technological solutions adopted.

TACS then gave way to the digital GSM system specified by ETSI, the organisation whose mandate is to develop European telecommunications standards. Figure 2.2 shows how the massive upswing in the number of digital system users in Europe began in 1994, while the total number of analog users started to drop in 1997. Interestingly, the breakdown between analog users and digital users differs significantly from country to country. This is due to the choices made by the operators: in all cases where an interim solution for transition to GSM was adopted (as was the case in the United Kingdom and Italy), growth began as early as the beginning of the '90s. In these countries, analog users still account for a far from insignificant share of the market today.

Today, service requirements are changing rapidly. As mentioned above, mobile data services – like their fixed network counterparts – will become increasingly important in the near future, and the signs of this trend are already apparent. The variety of envisaged services, together with the purely quantitative need for new spectral resources, have led the regulatory bodies, operators and manufacturers concerned to embark on a new, 'third-generation' stage of development, for which ITU has established the minimum service requirements.

As can be seen from Figure 2.3, the services envisaged for third-generation systems involve features which are not provided by current services. The major innovations regard the variety of capacities – or in other words, the bit rates – that are required, and the connectivity characteristics.

In addition to the classic point-to-point services, a significant rise in multicasting and broadcasting services is expected.

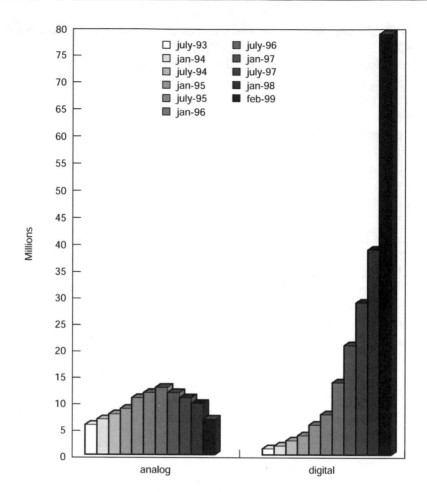

Figure 2.2   Users of analog and digital systems.

As can readily be imagined, such a large variety of services, together
with the general increase in required bit rates, will make it necessary
to identify new bands for third-generation services. The frequencies
assigned to GSM in the 900–1800 MHz band, in fact, can still cope
with a considerable increase in voice traffic, but certainly will not be
able to satisfy emerging demand for new services, and for data
services in particular. In addition, the GSM system itself is currently
developing in the direction of data services, which will shortly put
further pressure on spectral resources.

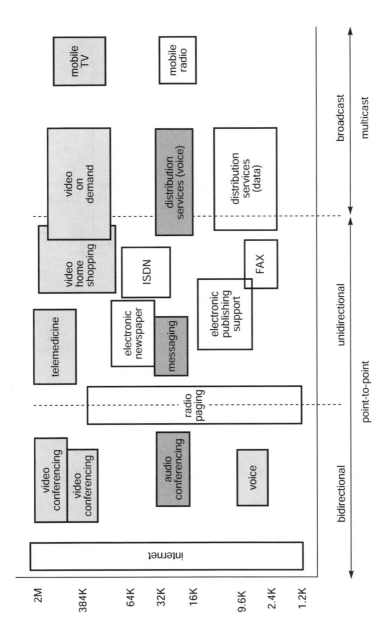

Figure 2.3 Innovative third-generation services.

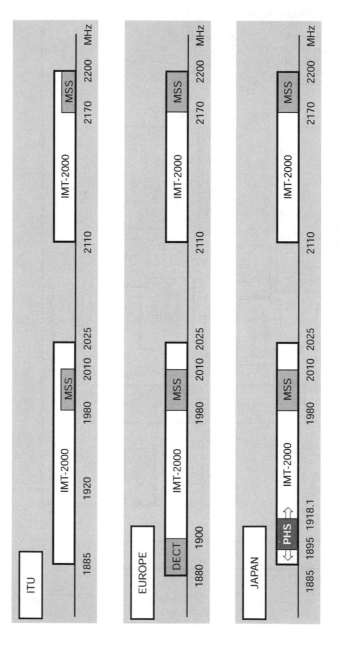

Figure 2.4 The band assigned to UMTS.

The WARC (*World Administrative Radio Conference*), which is responsible for assigning radio frequencies on a world-wide basis, acknowledged this need in 1992, when it assigned a band between 1885 and 2025 MHz and 2110–2200 MHz to third-generation systems. As can be seen from Figure 2.4, in Europe the first 15 MHz of the band coincide with part of the band currently used by the DECT (*Digital Enhanced Cordless Telecommunications*) system. The remaining portion of the spectrum for the terrestrial segment has been divided into a 'paired' part (60 + 60 MHz, from 1920 to 1980 MHz for the up-link and from 2110 to 2170 MHz for the down-link), i.e. with symmetry between the two links, and an 'unpaired' part (35 MHz, from 1900 to 1920 MHz and from 2010 to 2025 MHz), where there is no *a priori* distinction between the portions assigned to the up-link and down-link, respectively. Consequently, the bandwidth available for the terrestrial component of third-generation mobile systems in Europe is equal to 155 MHz. The 'paired' 1980–2010 MHz and 2170–2200 MHz bands have been assigned to the satellite segment for Region 1 (MSS).

The band allocated by the ITU and that characterised in accordance with European and Japanese needs are shown in Figure 2.4.

Other initiatives have been taken to promote the development of mobile communications towards third-generation systems and services, and to guarantee that the necessary resources are dedicated to them. The initiatives include the UMTS Forum, a free association promoted by operators, manufacturing industries, service providers and research centres with the aim of clarifying service, market and bandwidth requirements for UMTS. The association has devoted considerable attention to market assessments and to estimating the radio spectrum, which will be needed for third-generation systems.

In particular, the UMTS Forum has calculated what it regards as the 'optimal number' of operators in each country in the light of the band assigned by the WARC and of the minimum technical, economic and market conditions required to operate in the service conditions envisaged for the third generation. The Forum's assessment is based on the following assumptions:

- Between 2000 and 2005, the majority of innovative (multimedia) services will be assigned to UMTS.

- Only 10 percent of voice and low-capacity data services will be handled by UMTS (it is assumed that 90 percent of the traffic generated by these services will be absorbed by second-generation systems).

- The minimum channel spacing (or in other words, the minimum bandwidth that can be assigned to each operator) will be 5 MHz.

- The full 155 MHz bandwidth earmarked for Europe will be available for third-generation systems, as envisaged by WARC.

The assessment took into account the high degree of asymmetry that will characterise data services, both at the 'session' level (terminal-network transactions lasting a few minutes) and over longer time periods (an entire day). It is believed that the level of asymmetry (defined as the ratio of bits transmitted on the down-link to those transmitted on the up-link) may reach several dozen or even several hundred in the same session.

Assumptions regarding data service penetration put multimedia users in the year 2005 at 16 percent of the total, while penetration is expected to rise to 30 percent in 2010. Spectrum requirements were calculated for the high-density urban environment, and thus reflect the situation with most congestion. In the scenario thus outlined, the UMTS Forum has identified the minimum conditions for each operator (see Figure 2.5). On the basis of these projections, the minimum economical conditions for the service are such that there could be no more than four UMTS operators per country.

The third-generation (UMTS) application solutions will be made available through advanced terminals provided with such features as high-definition colour displays, paper-like display functions, or even integrated TV cameras. They will rely on the network's ability to guarantee high transport capacities by means of packet- and circuit-switching techniques. It is almost certain that UMTS features will vary from country to country.

In countries like Japan, the third generation will cover voice service bandwidth needs to a large extent. European countries will assign UMTS a more innovatory role, and will maintain most voice services on GSM. The driving application sectors will be:

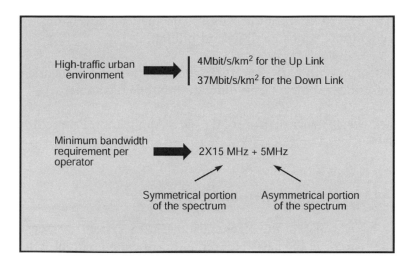

Figure 2.5   Minimum feasible conditions for UMTS according to the UMTS Forum.

- Interpersonal communication (audio and video telephony).

- Message services (Unified messaging, Video-Voice mail, Chat).

- Information distribution (Web browsing, information services, public services, tele-education).

- Enhanced positioning applications (personal navigation, driving support).

- Business services (process management, indoor mobility).

- Mass services (banking services, e-commerce, monitoring, help desk services).

The development of these application sectors will benefit from support platforms that provide basic elements which are located functionally on the network. These elements can be grouped in the following three categories:

- Platforms with tool-kits (tools for media conversion, speech recognition, traffic analysis and billing).

- Platforms with basic servers (for certification and authentication, subscriber profiles, positioning and mobile e-commerce).

- Intranet–Internet platform (Wireless Application Protocol).

2.3 Major system innovations

As we have seen, the third-generation UMTS system was designed essentially to meet the new requirements associated with mobile data services. The integration configurations, assigned band and the minimum capacities specified at the world-wide level must accommodate a number of factors, including:

- The universal nature of the services provided, achieved chiefly through direct and indirect interaction with the Internet world.

- The transport requirements (144 and 384 kbit/s and 2 Mbit/s, depending on the service environment and the mobility characteristics).

- The flexibility requirements, satisfied through two radio access modes, the widespread use of variable bit rate schemes, and the possibility of controlling significant levels of service asymmetry (i.e. differences in capacity between the up-link and down-link).

While the need to ensure universal service can be met through the network functions, the levels of integration (with the Internet, for example) and the development of intelligent functions, transport and flexibility requirements are to a large extent covered by the methods for accessing and using the radio resource chosen for third-generation systems.

By comparison with systems such as GSM, this area calls for a paradigm shift which is particularly significant precisely because of the variety of data traffic (service classes) and this traffic's changing bit rate characteristics.

Radio access, which is based on the CDMA (*Code Division Multiple Access*) technique and is described in Chapter 3, by nature accommodates a broad range of bit rates on the same (single) radio channel. The bit rate can even be varied during the same connection, if the source so requires. In TDMA (*Time Division Multiple Access*) systems such as GSM, transport capacity scalability is directly connected to the multiple use of the individual available channels, or bearers, and is thus less flexible.

For CDMA, moreover, the radio resource is used in a new operating setting, which avoids frequency plans and can be termed 'interference driven', in the sense that decisions to allocate the resource to a connection must deal directly with the level of interference present at the moment the decision is made. This level, in fact, is not guaranteed beforehand as it is in TDMA systems, where the frequency allocation plan (formulated at the planning stage) provides the necessary interference protection.

Solutions based on ATM (*Asynchronous Transfer Mode*) access are gaining creditability not only because of this technique's potential for success in many contexts such as fixed core networks and data network backbones, but also because of specific characteristics that make it particularly suitable for the UMTS access network.

Choosing ATM for the access network also encourages its extension to the UMTS core network, especially with an eye to avoiding double transcoding (as is now the case with GSM) for mobile-to-mobile voice traffic.

In the Core Network (the switching part of the UMTS network), it is expected that data services will be integrated directly with the IP network, with the latter's networking capabilities and with the variety of services that have already been developed for the fixed network.

Above the radio base station controller, the problem of dependency on the radio portion's characteristics has been entirely resolved, thus paving the way to diversified handling for voice traffic and data traffic.

Studies addressing the UMTS core network segment have resulted in a solution that can now be regarded as consolidated. In the initial stage, separation between voice and data will be maintained, while data will use essentially the same solution adopted for GPRS. Stage two will provide integration between the two types of traffic, which could be accomplished on a circuit-oriented platform based on ATM, or on a platform which is packet oriented and thus based on IP. The latter option appears to be particularly interesting, as it places us in the mainstream of integration between mobile telecommunications and information technology, with all of the advantages which such integration offers in terms of service accessibility.

2.3.1 The evolution of radio technology and the access network

In service requirements as in other areas, the last decade has brought significant advances in radio functions. Two access techniques have been selected for UMTS: code division multiple access (CDMA), and hybrid time division-code division multiple access (TD-CDMA). CDMA access belongs to the family of spread spectrum techniques, and is one of the major breakthroughs with respect to the access schemes used for today's mobile communications. A thorough analysis of the CDMA technique is provided in Chapter 3.

The new radio functions introduced by the CDMA technique and the service requirements illustrated above also bring important innovations in access network architecture for the new generation of systems. The general structure now being specified is illustrated in Figure 2.6.

The radio interface is specified in a framework in which the access functions are separate from the higher-layer functions. The latter typically deal with the direct relationship between the mobile terminal and the core network. The access layer provides a set of services to the higher layer through specific *Service Access Points*, or SAPs.

The approach embodies several of the basic concepts developed for third-generation system architectures. In particular, it draws on the idea of radio technology-dependent and radio technology-independent functions.

The signalling and control functions activated between the *Mobile Station* (MS) and the *Radio Access Network* (RAN) tend to be dependent on the radio technology, while those arising between the mobile station and the core network are typically independent of the radio technology, in the sense that they remain the same regardless of the access technique used.

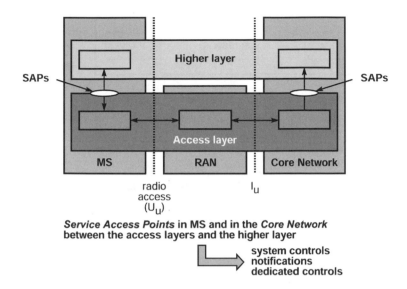

Figure 2.6 Reference architecture for the access network.

The fact that there is a single I_u interface between the access network and the core network reflects the ITU position for the IMT-2000 system model, which aims at network infrastructures that are capable of handling different radio access modes.

This satisfies the need for 'openness' on the part of the various entities, and for complete network flexibility as regards the variety of radio and network solutions and the associated compatibility requirements. The physical divisions between the various components correspond to the main interfaces: the U_u radio interface and the I_u interface between the Radio Access Network and the Core Network (Figure 2.6).

Normally, the access network performs the task of controlling the radio resources and the associated mobility functions, including hand-over control. The Core Network provides call control and performs mobility and high-level security functions such as location updating and authentication.

This division illustrates the *Generic Radio Access Network* concept, which ETSI has taken as the model for specifying UMTS network architecture.

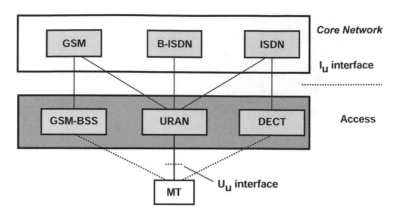

Figure 2.7   The ETSI view.

The Generic Radio Access Network as shown in Figure 2.7 can be integrated with a plurality of fixed networks, allowing for the latter's level of development and the stage reached in the evolution of the UMTS Radio Access Network. In the case considered in the figure, the UMTS terminal is assigned the ability to access a large number of Core Networks through the new U_u radio interface and the I_u network interface. Obviously, the terminal must be able to interwork directly or indirectly with the various Core Networks' network layers.

It should be noted that the UMTS terminal is thus able to provide services through GSM, as well as through the UMTS access network.

In the approach adopted for the Generic Radio Access Network, gradual UMTS coverage can be provided in several ways. If an operator decides to use the new system only in terms of radio band in the early stages of development, services will be made available through the current GSM network infrastructure. Conversely, if an operator decides to offer the innovative UMTS services right from the outset, the GSM infrastructure will immediately have to provide the necessary innovative functions. In both cases, full compatibility will be achieved through dual mode terminals, i.e. terminals capable of operating on both networks.

A more detailed view of the UMTS access network architecture is shown in Figure 2.8. At ETSI, this part of the network is called the UMTS *Terrestrial Radio Access Network*, or UTRAN. UTRAN consists

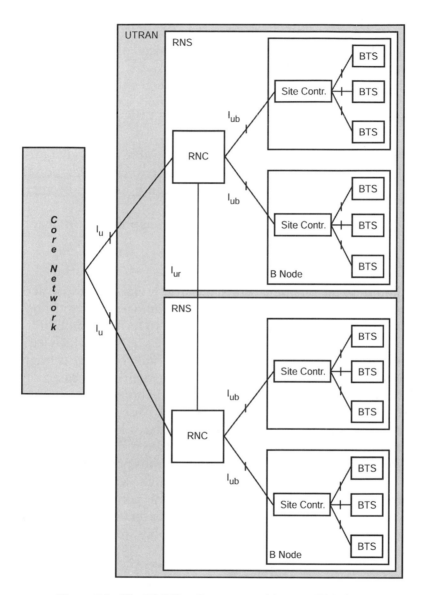

Figure 2.8   The UMTS radio access architecture (UTRAN).

of *Radio Network Subsystems* (RNSs) connected to the Core Network through the I_u interface. The Radio Network Subsystems can be interconnected directly with each other through the I_{ur} interface.

Each Radio Network Subsystem consists of a *Radio Network Controller* (RNC) and one or more *B Node*s. The latter entities are responsible for controlling radio resources in the cells assigned to them. A B node can contain one or more radio stations called *Base Transceiver Stations* (BTS), which are controlled by a site controller.

The Radio Network Controller is responsible for local control of hand-overs and of the functions associated with macrodiversity between different B nodes (the macrodiversity function is described in Chapter 3).

A discussion of the UTRAN access network is presented in Chapter 4.

2.3.2 The evolution of mobility control

In addition to the functions which are strictly associated with activating and releasing radio channels, a mobile radio network must be able to guarantee a series of capabilities linked to terminal mobility.

The most important of these capabilities is that which ensures connection continuity despite the fact that radio coverage is fragmented into cells. As is well known, mobile networks are based on the cellular concept, which was adopted in order to guarantee frequency reuse: the same frequency (radio channels) used in a given cell is reused in another cell provided that the latter is located at a distance sufficient to limit mutual interference between the two. For mobile networks, the concept of cellular coverage thus poses an interconnection problem which is entirely new, and unknown in fixed networks (where traffic sources do not change their point of access to the network over time): that of hand-over, or maintaining continuity for the current connection when the mobile terminal crosses cell borders.

In networks with TDMA access (or FDMA – *Frequency Division Multiple Access*), hand-over is accomplished by changing the frequencies of the connection which is cutting across cell profiles. In GSM, the terminal engaged in a call, continuously measures the level and quality of the signal received from the cells adjacent to the cell where it is currently located (it also measures the radio channel carrying the call in progress). This measurement is sent to the network on a channel associated with the one which is active between the network and the terminal.

As soon as the quality of the communication in progress drops below a certain threshold, the network decides to change the channel in current use with another belonging to another base station, and which the measurement indicates is received best by the mobile terminal. A procedure is thus initiated which changes the communication channel in real time, guaranteeing continuity for the call in progress. Obviously, the call is re-routed on the network at the time this change is made.

For CDMA, the typical distance for frequency reuse is 1: all terminals, in fact, use the same radio frequency. This also brings about a substantial innovation in the methods used to interconnect the terminal with the network, and in the functions used to maintain connection continuity. Given its characteristics, a CDMA system can provide fully asynchronous hand-over functions. This capability, called *soft hand-over*, is made possible by the macrodiversity mechanism. Macrodiversity makes it possible to maintain a connection between the mobile terminal and the network active over several radio links, which can also be activated through different radio base stations. This means that the same user information is carried over the radio interface through the (several) channels which are active with the various stations.

The set of active stations – called the *active set* – takes charge of the connection. Soft hand-over and macrodiversity are thus closely connected with each other. In the 'channel changing' stage typical of the hand-over process, connection continuity is guaranteed through the multiple paths set up between the mobile terminal and the 'controlling' point in the network. The macrodiversity mechanism relieves the stringent delay constraints that affect the hand-over process in many TDMA systems (in GSM, for example, channel changing takes place by means of a 'hop' between the radio channels of the two cells involved in the process, and hence through an 'interruption' whose duration must necessarily be limited).

The other procedure associated with mobility is that of mobile terminal locating, which enables the network to recognise where

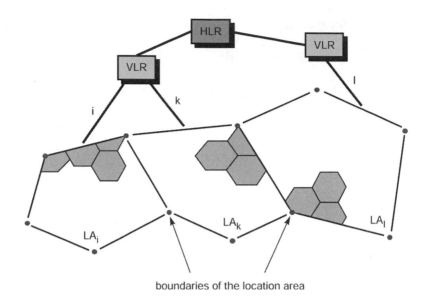

boundaries of the location area

Figure 2.9   GSM location updating.

the terminal is located instant by instant. For this purpose, cells are grouped in *Location Areas*. Whenever the mobile terminal crosses the boundaries of a location area (see Figure 2.9), the network registers this change by updating the databases, which thus follow the mobile terminal's movement in the network (with a granularity corresponding to the Location Area). The *Home Location Register* (HLR) contains all user data (e.g. billing, security and service data), while the *Visitor Location Register* (VLR) contains the data needed locally at the exchange in order to manage mobile calls. In other words, the VLR contains a subset of the HRL data.

When it changes Location Area, the mobile terminal initiates a procedure which updates the status of a pointer in the HLR to indicate its new position, i.e. the new location area. The same procedure ensures that the user's local data are transferred to the new Visitor Location Register. There is a correspondence between registers and Location Areas.

In current networks, the Location Area coincides with the area in which the paging message is sent for an incoming call, i.e. a call

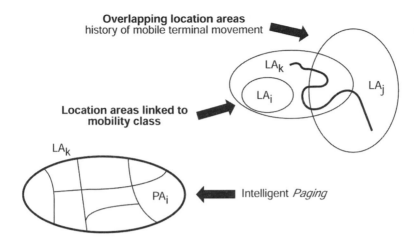

Figure 2.10 Innovations in paging and locating procedures.

directed to the mobile terminal. In advanced mobile systems, specifying the Location Areas and paging areas will be one of the more
critical topics. This is because of the microcellular structure of
radio coverage, and the variety of services and mobility levels.
New mobility management mechanisms are now being investigated,
which will be able to reduce control traffic and radio resource
occupation.

One such mechanism (Figure 2.10) consists of introducing over-
lapping Location Areas. At the cost of a minor increase in the mobile
terminal's processing load, this solution can reduce the number of
updates, given that it introduces a sort of hysteresis in the updating
process.

The variety of traffic and mobility levels may make it advisable to
specify Location Areas as a function of user class. This second solution (Figure 2.10) also reduces update frequency, again with a small
increase in user information content.

A third solution (also shown in Figure 2.10) introduces paging
areas which do not coincide with the Location Areas. Paging for an
incoming call starts in the paging area where the mobile terminal is
most likely to be found. (e.g. the area where it received or generated
the last call). Paging is then extended to the entire Location Area only

if the network receives no response. This mechanism is useful for user classes with a low level of mobility, where the savings thus gained in radio resources is not offset by a substantial worsening in average call activation time.

2.3.3 Architecture and core network evolution

In the context described above, several technological and service factors can play an important role in framing architectural decisions. One of the first technological factors is the ATM technique which, as we have said, can provide enormous advantages if used as a transport technique in the UMTS infrastructure.

The major benefits include: efficiency, flexibility, and the ability to monitor quality of service. Several aspects are particularly important for mobile communications:

- The ability to multiplex data streams statistically.

- Dynamic bandwidth allocation.

- The possibility of avoiding double transcoding or double multiplexing for local mobile-to-mobile connections.

- The ease with which alternative routing solutions can be used in the event of faults.

However, using the ATM technique involves a number of problems because of the substantial difference between the size of the data blocks exchanged on the radio access and that of the ATM cells. Typical dimensions, in fact, are around 100 bits for the radio interface, whereas an ATM cell carries 48 bytes (i.e. 384 bits) of information. To achieve high efficiency, many radio blocks would have to be accumulated in a single ATM cell, which would cause an unacceptable latency for delay-sensitive services.

The solution may lie in adopting the ATM *Adaptation Layer* 2, specified by the ATM Forum and by ITU SG13. This adaptation layer makes it possible to multiplex several user connections on a single ATM virtual connection.

A header must be added to each block in the ATM cell in order to identify the associated connection and hence the particular origin-destination relationship. The *Channel Identifier* (CID) performs this role, and also defines the block's length. The CID is used in the transit node to ensure that the different streams which were previously multiplexed on the same virtual connection are routed correctly.

Another factor associated with the architecture, which is of importance to technological innovation, is the spread of new services. As mentioned earlier, increasing interest in mobile data services is fuelled by the rapid rise in personal computer ownership and the growing success of Internet-based services. This is one of the aspects which is most emblematic of the convergence between the two worlds in which different network and service models have taken root: that of telecommunications and that of information technology. The two alternatives which characterise them hinge on how intelligence is distributed and, until not so many years ago, had progressed along entirely separate courses.

While the telephone network developed all of the functions associated with communication between any two of its terminations (network intelligence), computer interconnection needs have relegated the network to performing a simple transport function, moving intelligence 'to the edges', i.e. to the very terminations – the computers – that the network connects.

Another basic difference between the two contests consists in the methods used to make connections between the network terminations. Each interconnection, which entails a direct voice or video interaction between connected entities, normally calls for substantial and continuous exchange of data between the two entities, while transmission modes must satisfy stringent constraints on the maximum permissible delay in transferring information (and on the delay's probability distribution). By contrast, interconnection between computers involves the need for *block* exchanges, where transfer delay is much less vital than the transfer's correctness, or freedom from errors. This need is joined by another: that of being able to rely on a single communication method whereby computers with dissimilar processing approaches, operating methods and capacities can dialog with each other. In a packet-switched network, the communication subnetwork does precisely this, guaranteeing transfer uniformity and error protection mechanisms while providing dialog modes that cut across differences.

Today, the spread of personal computers, the Internet, and World Wide Web navigation traffic are the clearest signs of how successful this type of information technology model has been. The Internet supplies the routing and transport functions needed in a widespread computer network. The most commonly used protocol is the *Internet Protocol* (IP), which guarantees that packets are routed between the network's nodes to their final destination. Each packet is switched in accordance with a mechanism that assigns routing decisions to each node involved on the basis of the information content included in the packet's header.

The IP nodes (the routers) can be interconnected through heterogeneous subnetworks such as telephone or local area networks. The computers (hosts) are connected through suitable IP addresses, which assume the mnemonic form of the terminations (i.e. personal computers or corporate sites). Specifically, the IP address is made up of two components, for the network and for the host:

- The first component contains all of the elements needed to route the packet towards the destination network.

- The second component makes it possible to deliver the packet to the final computer in the destination network.

Two protocols are used at the point-to-point transport level: the *Transmission Control Protocol* (TCP) and the *User Datagram Protocol* (UDP), which guarantee correct transmission between the origin and destination nodes for connection-oriented and connectionless services, respectively. The main services provided through the Internet include: file transfer, remote log-in, and file and information searches.

More recently, IP services have also been made available when the terminal computer is moved away from its usual point of access to the network (Mobile IP). Making the service available 'on the move' is particularly difficult to guarantee whenever a new access point belongs to a network other than that on which the terminal is resident: in this case, both of the address components associated with the terminal must change in some way.

It is thus necessary to specify a new protocol capable of coping with such situations. In current IP protocols (version 4), the *Mobile IP* function is performed through an encapsulation mechanism called *tunnelling*. Packets sent by a network user to the *Mobile Host*, which has moved from its own network to visit another one (the *Foreign Network*), are encapsulated by the *Home Agent* (so called because it is located on the home network) and transmitted with the address of the *Foreign Agent* on the foreign network which depacketises the data and delivers them to destination. In the opposite direction, packets travel directly from the Mobile Host to the network termination connected to it. This procedure requires that the *Roaming Mobile Host* first register itself on the foreign network, and that the Foreign Agent inform the Home Agent of this registration, providing the latter with all information needed to re-route packets (Figure 2.11).

However, version 4 of the IP protocols has a number of limitations which are now being overcome through the introduction of an updated version: version 6. In this new version:

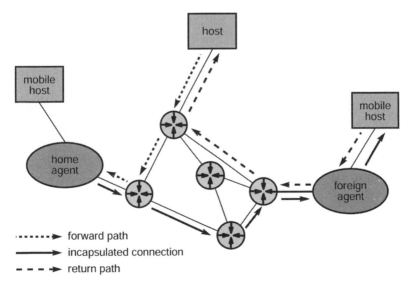

Figure 2.11   Mobile IP.

- Addressing potential has been increased by introducing a special field for the renumbering required for *Mobile IP*.

- A highly functional routing mechanism has been introduced which prevents network overloading induced by the presence of many computers operating outside of their home networks.

Note that locating requirements are equivalent to those typical of mobile networks. The functions carried out by the Agents are similar to those performed by the Location Registers in any given mobile network or between different mobile networks.

Today, the spread of data services (chiefly those based on IP) is increasingly interwoven with the spread of the concept of mobility. Clearly, in fact, the burgeoning growth of mobile communications is increasing marked – and not just in the forecasts – by keener interest in multimedia mobile communications, where data, voice and image connections can coexist in the same call.

In this sense, the *General Packet Radio Service* (GPRS) introduced in the GSM system can be regarded as an intermediate step towards UMTS. As such, it is an example of integration between circuit-switched functions and packet switching. The main problem which has had to be solved for the GPRS is associated with the compatibility between an environment which is heavily oriented towards circuit switching (as used in the GSM access network) and the requirements typical of packet switching. To overcome this problem, the GPRS has been specified through the adoption of interworking functions between the GPRS support nodes (SGSN – *Serving GPRS Support Node* and GGSN – *Gateway GPRS Support Node*) and the GSM radio access network (Figure 2.12).

This interworking takes place between the Serving GPRS Support Node and the GSM *Base Station Controller* (BSC). The A-bis interface between the GSM *Base Transceiver Station* (BTS) and BSC has been retained as the common interface for the circuit and packet switched modes. As in GSM, identification between the connection and the TDMA frame time slot is maintained as far as the BSC level.

The identification scheme described above is no longer necessary with CDMA access, where variable bit rate mechanisms will be exten-

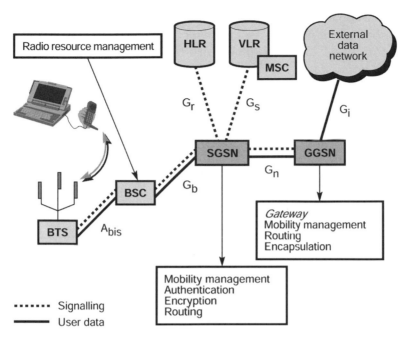

Figure 2.12   The GPRS system.

sively used in the radio interface. For this reason, the choices made in this connection will be thoroughly overhauled for third-generation systems, not least through the adoption of IP type protocols.

The first aspects of architectural innovation will come into play during the transition between second- and third-generation systems.

Figure 2.13 illustrates one of the paths which evolution from GSM (GSM phase 2+ in this case) to UMTS could take. Here, the GSM system is seen as operating in a two-fold capacity: on the one hand, it provides classic circuit-switched services (voice and low bit rate data) through the infrastructure based on the *Mobile Switching Centers* (MSC) and *Gateway MSC*, while on the other hand it provides packet-switched data services by means of the GPRS system described above.

The configuration shown in the figure could be the first step towards UMTS, a step which chiefly solves the capacity problems. Here, in fact, the innovative radio access is used simply to increase the available capacity, or in other words the spectrum. The configuration does not make it possible to supply services which are innovative with respect to GSM, as the fixed network functions remain those of GSM and are not updated for the new service requirements. Compatibility between UMTS access and the GSM Core Network is guaranteed by the interworking functions for the two phase 2+ voice and data segments.

Using the innovative radio access in the approach shown in Figure 2.13 can also be of interest to consolidated GSM operators, e.g. in cases where broadband data services fail to provide the essential boost needed for short-term growth in mobile communications. Substantially, the architecture would confirm the present-day service set-up, based on a preponderance of voice and low bit rate applications.

The solution shown in Figure 2.14, on the other hand, is an answer to the need to provide innovative data services (and innovative bit rates) in the short term. The new UMTS infrastructure is integrated

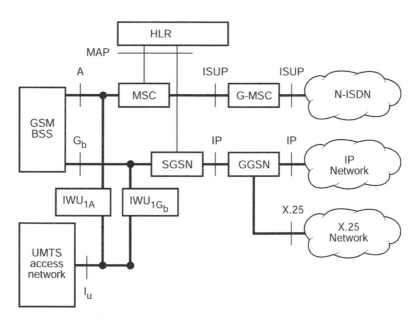

Figure 2.13   Capacity-oriented GSM-UMTS migration.

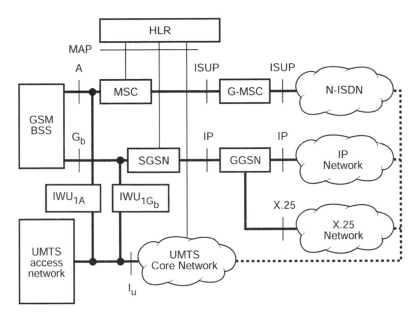

Figure 2.14 Service-oriented GSM-UMTS migration.

with GSM. The latter maintains its dual voice-data character (GPRS) even in cases where a UMTS access is provided through the same interworking functions discussed above. The UMTS Core Network performs circuit and packet switching functions.

The UMTS Core Network is described in Chapter 5, where the concepts which have been briefly outlined here are discussed in depth.

3

UMTS System Radio Access

Sergio Barberis, Bruno Melis and
Giovanni Romano

In mobile radio systems, the user signals are transmitted on a radio
carrier. As the electromagnetic spectrum available to the various
operators is limited, it is important to ensure that the radio
resource is utilised as efficiently as possible. This kind of radio
resource management is accomplished using multiple access techniques. With these techniques, the shared resource (the spectrum) can
be divided shared among several users, ensuring the quality of
service.

The procedures used to transmit and receive user signals (radio
access) are an integral part of access techniques. The OSI (Open
System Interconnection) protocol stack physical layer establishes
the procedures whereby the radio medium is accessed. The physical
layer protocols plus the OSI stack layer two and three protocols
constitute the radio interface for the UMTS system.

This chapter will describe the access techniques used and the main
OSI stack physical layer characteristics specified for the UMTS
system. In particular, given that UMTS's objective is to provide multi-

media services, attention will focus on those characteristics which make it possible to support innovative applications.

The process of specifying the UMTS radio access got under way at ETSI in 1997, with the creation of several working groups who set out to develop the radio solutions presented by various companies.

ETSI examined four basic alternatives for implementing the UMTS radio interface: the W-CDMA (*Wideband-Code Division Multiple Access*) technique, the TD-CDMA (*Time Division-Code Division Multiple Access*) technique, a technique based on time division transmission called W-TDMA (*Wideband-Time Division Multiple Access*), which is similar to that used by GSM but provides a much higher transmission rate, and a multi-carrier technique called OFDMA (*Orthogonal Frequency Division Multiple Access*). By the summer of 1997, however, it was clear the only the first two solutions had properties which were appropriate for the new system to be developed.

In January 1998, ETSI reached an agreement concerning the radio access technique to be used for UMTS. This solution, named UTRA (*UMTS Terrestrial Radio Access*), is based on the two W-CDMA and TD-CDMA proposals.

Specifically, the ETSI decision envisaged that:

- The system will use the W-CDMA access technique in the paired bands (with FDD duplexing).

- The system will use the TD-CDMA access technique in the unpaired bands (with TDD duplexing).

- Radio access specifications must be such as to ensure that low-cost terminals can be developed, while at the same time guaranteeing dual mode UMTS/GSM and FDD/TDD terminals.

- The FDD component will enable an operator to supply UMTS services with a minimum bandwidth allocation of 2×5 MHz.

The resolution that assigns bands to UMTS does not establish the duplexing method (TDD or FDD), which depends on the radio solution chosen for the system. It goes without saying, however, that the method will draw on solutions that have already been adopted in existing systems, such as the FDD approach when there are two separate and symmetrical bands for the two links (as in GSM, for example), and TDD when a single slice of the band is assigned to the system (as is the case with DECT).

In the FDD, transmission between the mobile terminal and the base station (on the up-link) takes place in one sub-band (generally the lower), while transmission between the base station and the mobile terminal (the down-link) takes place in the other sub-band. The transmission and reception operations can thus occur simultaneously, as the two signals are separated in frequency. This kind of transmission is particularly suitable for symmetrical services, where user information is transferred at the same rate on both links.

In the TDD component, the same sub-band is used for transmission on the up-link and on the down-link. In this case, the transmission and reception operations are separate in time. The moment of switchover between transmission and reception can be selected in such a way as to provide asymmetrical services, where user information can be transferred at highly dissimilar rates on the two links. An example of this type of service is the access to a database, where enormous high quantities of information are read after making brief queries.

3.1 The W-CDMA access technique

Radio systems transmit and receive on a common resource: the fraction of the electromagnetic spectrum assigned to the system by the regulatory bodies. In general, the fact that several users of the same system employ a shared resource, leads to situations of conflict if two or more users transmit at the same time and on the same frequency without taking special precautions. Thus, multiple access techniques have been developed in order to resolve possible interference between users and maximise the system's capacity, or in other words, the number of users that can be served with a predetermined quality of service.

Classic access techniques attempt to divide transmission resources, i.e. frequency (or bandwidth) and time as efficiently as possible between the users who wish to access the service. These techniques are called FDMA (*Frequency Division Multiple Access*) and TDMA (*Time Division Multiple Access*), respectively.

The FDMA technique consists of dividing the bandwidth assigned to the system into a certain number of slices called 'channels' centred on a carrier frequency. With this technique, the basic resource is thus the radio carrier. Each user is assigned to a channel (i.e. a carrier) for the entire duration of the conversation. Other users can access the channel after the first user's conversation has ended. This technique is used by first-generation analog cellular systems such as TACS (*Total Access Communication System*) and AMPS (*Advanced Mobile Phone Service*).

The TDMA technique breaks up the transmission resource into time fractions called 'time slots'. Multiple users can make use of the band assigned to the communication at different moments, or slots. In this case, the basic resource is the time slot assigned to the communication. In general, mixed TDMA-FDMA techniques are used, where the bandwidth assigned to an operator is divided among different FDMA carriers, each of which is shared by the users with the TDMA technique. The basic resource is thus the time slot–radio carrier pair. The TDMA/FDMA technique is used by second-generation digital cellular systems such as GSM and PDC (*Personal Digital Cellular*).

The CDMA technique differs from the methods described above, in that it enables users to transmit at the same frequency and in the same instant. Users are separated by assigning a different 'code' (or sequence) to each. Through the sequences, the user data to be transmitted is coded on a unique basis so that it can be distinguished from that of the other users. In the technical jargon, this operation is known as *spreading*. The basic resource is the sequence associated with each user signal. Here again, hybrid approaches with the techniques described above can be implemented. Depending on which techniques are associated, the basic resource is identified by the set of the parameters involved.

In the spreading operation, each signal to be transmitted on the radio channel is associated through a multiplication operation with a numerical sequence or code whose transmission rate, or *chip rate*, is much higher than the rate used for the information to be transmitted.

Again in the technical jargon, the bits obtained after this operation are called *chips*. The code sequences assigned to the users who share the same channel differ, and are selected so that there is very little correlation between them. Consequently, under ideal conditions, the reverse operation in reception – *despreading* – will cancel out the effects of mutual interference and make it possible to extract the desired signal.

Under real propagation conditions, the distortions and interference which the signals undergo along the communication channel result in degraded orthogonality conditions, so that the number of signals that can be superimposed on the same channel is limited. The system's capacity limit is thus given by the level of interference remaining after the despreading operation. Minimising this residual interference level is thus of fundamental importance.

The bandwidth occupied by the transmitted signal is clearly greater than that which would be strictly necessary in order to transmit the information. In reality, the apparent loss of spectral efficiency is compensated by the ability to superimpose several signals on the same radio channel. The higher the ratio of chip rate to user information bit rate, the greater the interference robustness – and thus the number of users who can transmit simultaneously on the same channel – will be. Interference robustness is so high that the same carrier frequency can be used in all cells of a mobile radio network.

The CDMA technique can be explained through a simple analogy: let's assume we are in a conference room where three speakers are simultaneously making a presentation, each in a different language, English, French and Italian, for instance. A hypothetical listener in the audience who knows only English will be able to follow and understand the presentation in English to a certain extent, while the presentations in other languages will be perceived as background noise (see Figure 3.1). The same thing happens in a CDMA system, where the information sequence 'encrypted' with the code used by the receiver is recovered, while the sequences using other codes will be cancelled or, in real conditions, highly attenuated.

The three techniques for access in the time-frequency-power domain are shown schematically in Figure 3.2.

Figure 3.1   Analogy used to explain the CDMA technique.

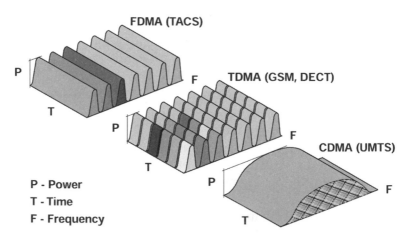

Figure 3.2   Basic access techniques in the time-frequency-power domain.

Close-up 3.1 – Spreading and despreading

With the CDMA technique, the transmitted signal's bandwidth is considerably larger than that would be strictly necessary.

The effect of spreading on bandwidth is illustrated graphically in Figure 3.3, where $b(t)$ is the data signal, $c(t)$ is the code assigned to the user, and $B(f)$ and $C(f)$ are the associated power spectral densities. The symbol '*' represents the convolution operation between functions.

It is important to bear in mind that the bandwidth can be spread using information repetition techniques and error correction codes, as well as by using sequences with good orthogonality properties, which differentiate between among the various user signals.

Quantitatively, the overall increase in bandwidth is equal to the Processing Gain (P_G), which is defined as the ratio of the transmitted signal bandwidth (f_C) to the data signal bandwidth (f_b).

$$P_G = \frac{f_C}{f_b} \qquad (3.1)$$

As f_C is generally much higher than f_b, the bandwidth may be increased. The processing gain P_G may be in the range from by a few units up to some hundred times higher.

The number of chips representing a given bit that feeds the spreading block is defined as the *Spreading Factor*. In the UMTS system, this factor coincides with the length of the sequence used to distinguish between among the different user signals.

The difference between the processing gain and the spreading factor must be emphasised. The former includes all processing operations between the data source and the transmitting antenna which contribute to spreading the bandwidth. Error correction codes, for example, are included in the processing gain. The spreading factor, on the other hand, includes only the spreading operation itself, or, in other words, the multiplication of the signal by a sequence used to distinguish the user. The processing gain is linked to the CDMA technique's ability to reduce interference (its *interference rejection* capability), while the spreading factor is linked to the number of available sequences, and thus governs the number of users who can be served.

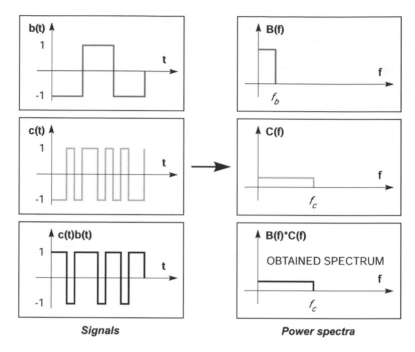

Figure 3.3   Increase in bandwidth as a result of spreading.

To recover the information-bearing signal at the receiver, the received signal is multiplied by the same code $c(t)$ assigned to the user: this is called the *despreading* operation. By means of a low-pass filter, the signal's information-bearing component is then selected. In order to perform the despreading operation, the code must thus be known to the receiver in some way: for example, the mobile terminal may be notified of the code via connection negotiation stage signalling. In addition, the code $c(t)$ applied in reception must be synchronous with that used in transmission.

Let us now suppose that jamming occurs, with narrow band interference superimposed on the transmitted signal. The despreading operation makes it possible to remove the code and thus recover the information.

At the same time, however, the interference is multiplied by a

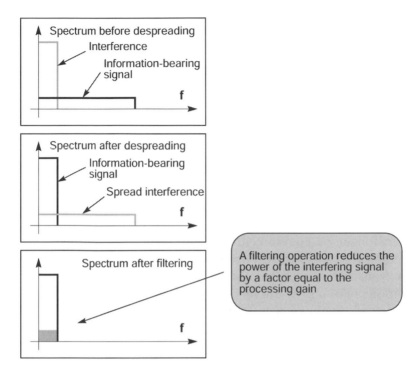

Figure 3.4 Interference rejection with the CDMA technique.

high bit chip rate sequence, which causes a bandwidth spreading and a power spectral density reduction. By means of a low-pass filter (with bandwidth f_b), the information-bearing signal is recovered, and the power of the interfering signal is reduced by an amount equal to the processing gain. This concept is illustrated graphically in Figure 3.4.

The example given above also applies in the case of wide band interference. For example, we can assume that the interference is the signal from a second user to whom a code $c'(t)$ differing from code $c(t)$ has been assigned. After the despreading operation has been carried out, the interference remains spread because the product $c'(t)\,c(t)$ is still a sequence with spread band f_C. As before, the filtering operation eliminates the interference signal component which falls outside of the useful band.

3.1.1 Capacity of CDMA systems

As will be recalled, the CDMA technique, in which the different signals are distinguished by different codes, in theory ensures that interference between the various signals is completely eliminated. In practice, however, propagation conditions reduce the orthogonality properties of the sequences used. As a result, the number of conversations is limited by mutual interference.

Unlike first- and second-generation systems, there is no predetermined limitation on resources, i.e. on the number of carriers or time slots: each time a new call is accepted, the quality provided to all the active users present on the system drops slightly. New calls can be accepted until the interference level is such that the desired quality can no longer be guaranteed. Nothing prevents a number of calls exceeding that established purely on the basis of quality considerations from being accepted for a limited period under emergency conditions. For all of these reasons, CDMA systems are said to provide *soft degradation* in quality. The capacity limitation due to interference can also be explained by an example similar to the analogy illustrated in Figure 3.1. Even if each pair of people are talking to each other in a normal tone of voice, the background noise can be loud enough to make it difficult to understand the words, if the number of separate conversations increases beyond certain limits.

In the example shown in Figure 3.1, all of the couples who are talking to each other in different languages (codes) will be able to continue their conversation providing that everyone speaks in a normal tone of voice. To make themselves heard better, one couple could decide to start shouting. In this way, however, they would cause a disturbance – or in other words, interference – which would prevent everyone else from continuing the communication. Hence, it is clear that the use of more power than necessary has the practical effect of annulling the cell's capacity.

A practical example of this phenomenon is provided by the case in which a mobile terminal is located very close to the base station. If appropriate measures are not taken and the mobile terminal transmits at its maximum power (or 'shouts'), all of the other users' signals which are received with a lower power because of their greater distance cannot be received correctly. This problem is known in the literature as the *near–far effect*.

In general, it is essential that all signals in the up-link for a given service are received at the radio base station with an equal power level. This can be achieved by means of a mechanism known as power control, whose purpose is to regulate the different signals' transmission level. To return again to our shouting couple, we can say that the principle effect of poor power control is to cause a significant reduction in capacity.

Power control is important both on the up-link and on the down-link. Nevertheless, there can be no doubt that it is more critical on the up-link, where many users transmit at the same time to a single base station. In this case, the interference due to users of the same cell depends on their distance from the radio base station. On the down-link, power control is less critical since there is a single transmitter (the base station) which bundles the various user signals and transmits them simultaneously to the different mobile terminals.

From all of these considerations, it is clear that any assessment of a CDMA system's capacity must be based on a careful evaluation of the system's interference level. A parameter of fundamental importance is the carrier-to-interferer ratio (C/I) of the service whose capacity has to be assessed.

3.1.2 Up-link capacity

Take the case of an isolated cell with perfect power control (all signals are received at the same power) in a single service scenario. Let N be the number of users, and C the effective power value of the reference communication. The remaining $(N - 1)$ users will contribute with an interference value equal to $C(N - 1)$. The C/I ratio is thus:

$$\frac{C}{I} = \frac{C}{C(N - 1)} = \frac{1}{N - 1} \tag{3.2}$$

In order to guarantee the desired quality of service, it is necessary to guarantee a given signal-to-noise ratio. The ratio which is generally used for this purpose is E_b/I_0, defined as the ratio of energy per information bit to interference spectral density (assuming that thermal noise is negligible). This can be expressed as follows:

$$\frac{E_{\mathrm{b}}}{I_0} = \frac{C/R}{I/W} = \frac{W}{R}\frac{C}{I} = \frac{W}{R}\frac{1}{N-1} \tag{3.3}$$

where W is the signal chip rate and R is the information source bit rate. As will be recalled, the W/R ratio is the processing gain.

The number of users who can be served in the cell is:

$$N \cong \frac{W}{R}\frac{1}{\dfrac{E_{\mathrm{b}}}{I_0}} \tag{3.4}$$

As can be seen, the number of users is influenced by two factors: the processing gain and the E_{b}/I_0 ratio. In particular, any technique that makes it possible to lower the E_{b}/I_0 ratio automatically entails a gain in terms of capacity.

Bearing in mind that the spreading process consists of multiplying the binary data signal by the signal associated with the spreading code, the radiated power will theoretically be zero during the source pauses. Consequently, with d the source average activity factor, the capacity value is increased by a factor $1/d$. In the case of voice trans-mission, a typical value is $d = 0.38$. In addition, if we assume that a sectorial antenna is used instead of an omni-directional antenna, the interference intercepted by one sector will be equal to $1/G_{\mathrm{s}}$ (G_{s} being the number of sectors or 'sectorisation gain'). The set of geographi-cally co-located cells, each of which is served by a sectorial antenna, is referred to as a cell site. The capacity of a cell site divided into G_{s} sectors is thus:

$$N \cong \frac{W}{R}\frac{1}{E_{\mathrm{b}}/I_0}\frac{1}{d}G_{\mathrm{s}} \tag{3.5}$$

Now take the case of multicellular coverage in which users are uniformly distributed and all require the same service. The effect of inter-cell interference is taken into account by increasing the inter-ference value by a factor (typically designated indicated as f or i) defined as the ratio of the interference received from the other cells to that produced inside the cell to which the mobile terminal is connected. Consequently, the capacity is reduced by a factor (some-times referred to as the CDMA system frequency reuse factor) equal to $(1 + f)$.

$$N \cong \frac{W}{R} \frac{1}{E_b/I_0} \frac{1}{d} G_s \frac{1}{1+f} \tag{3.6}$$

In this expression, the most critical element is the factor f, which is not readily determined. Values of this factor are typically 0.5–0.6 in the macrocell environment, though they vary according to the operating area.

The expression thus obtained can be generalised to cover the case in which users with different types of service are present. In any event, it should be noted that the formula given above is approximate, and is based on average values. A more accurate assessment must be based on a statistical analysis that takes all of the system's significant random components into account. In general, however, simulation techniques are required in order to obtain more precise information.

3.1.3 Down-link capacity

Though the down-link's capacity is also determined on the basis of the calculated C/I ratio, it is not readily translated into approximate formulas. The basic concepts associated with the down-link can be summarised as follows:

- A single source transmits towards many receivers. Interference is received by a few concentrated, higher-intensity sources (the radio base stations), rather than by a large number of mobile terminals scattered over a large geographical area.
- The properties of the spreading sequences significantly reduce (and in theory should eliminate) the interference generated inside the cell.

Capacity is thus calculated by assessing the carrier-to-interferer ratio for a generic user and checking that the serving base station has enough power to guarantee the required quality of service.

3.2 The TD-CDMA access technique

The TD-CDMA solution is a combination of a time division access technique as used in GSM and a code division technique such as W-CDMA.

In the time division technique, transmission over every radio carrier is organised in frames divided into transmission intervals called time slots. Unlike GSM, each time slot is not dedicated to a particular link, but can be used simultaneously by several different links, superimposing the signals in them by means of a code division technique similar to that described above. The theoretical maximum number of channels that can be multiplexed on each radio carrier is given by the number of time slots multiplied by the available number of codes.

In this system, the ratio of chip rate to information bit rate on the individual channels is fixed on the down-link and variable on the up-link. In order to set up channels with different capacities, it is thus necessary to bundle several elementary channels, up to the limit at which all channels available for a given carrier can be used simultaneously.

TDD transmission and services with asymmetrical bit rates (i.e. with a higher rate in one of the two transmission directions) can be provided simply by dedicating part of the time slots to one transmission direction and the remaining time slots to the opposite direction.

3.3 The radio interface

The UMTS radio interface, called UTRAN (UTRA Network), has been defined by the RAN specifications committee as part of a the 3rd Generation Partnership Project, or (3GPP). In particular, the 25.1 series specifications cover the performance requirements for the radio interface, while the 25.2 series covers specifications for the physical layer.

The main characteristics of the radio interface are summarised in Table 3.1. The parameters shown in Table 3.1 are described in detail in the following sections. At this point, however, it should be noted that the maximum 5 MHz channel spacing was introduced in order to satisfy the requirements deriving from ETSI's decision to operate with

Table 3.1 Main UTRA parameters

	UTRA/FDD	UTRA/TDD
Access technique	W-CDMA	TD-CDMA
Chip rate	3.84 Mchip/s	3.84 Mchip/s
Channel spacing	4.4–5 MHz	4.4–5 MHz
Frame duration	10 ms	10 ms
Slots per frame	15	15
Modulation	*Down-link*: QPSK; *up-link*: dual code BPSK	QPSK
Reception	Coherent	Coherent
Rate	Variable (every 10 ms). Different rates can be achieved by changing the *spreading factor*, by assigning several codes to the signal to be transmitted, or (for TDD only) by bundling several *time slots*	

a bandwidth of 2 × 5 MHz. Though other chip rates (which call for different channel spacings) are envisaged at the moment, specifications for these values have not yet been developed. Consequently, they can be regarded as part of the system's possible future evolution.

As required by the ETSI decision, and in order to facilitate the development of dual mode FDD/TDD terminals, the parameters of the two components have been harmonised to the greatest possible extent. As a result, the specifications for the two components often consist of the same elements. The basic characteristics of the FDD component will be described below, indicating the differences with respect to TDD.

3.3.1 Correspondence between transport channels and physical channels

The UTRAN physical layer provides services to the higher layers, transmitting on the physical carrier information generated starting from the OSI stack Layer 2.

Specifically, the transport channels are the services which the physical layer provides to the higher layers. Transport channels are defined on the basis of the type of information they transfer, and how it is transferred on the radio interface. They can be grouped into two classes: common channels (where information is transmitted to all mobile terminals without distinction), and dedicated channels, where communication takes place towards a single terminal by associating it with a physical channel, i.e. a code and a frequency or, in the case of TDD, also a time slot.

As will be noted, the number of transport channels is much higher than for GSM. This is due to the fact that the UMTS system has not been optimised for a single service (voice), but must on the contrary be capable of simultaneously providing services with highly dissimilar service characteristics.

Dedicated channels

A single type of dedicated channel is envisaged, and is thus called the Dedicated Channel (DCH). It is used on both the up-link and the down-link to transport user and control information between the mobile terminal and the network.

For the TDD component, the possibility is envisaged of having a channel dedicated to re-transmitting user information from one network element to another (with the mobile terminal operating as a relay). This is the ODMA Dedicated Transport Channel (ODCH).

Close-up 3.2 – The ODMA technique

It is envisaged that the Opportunity Driven Multiple Access (ODMA) technique will be adopted in the TDD component. With this technique, the mobile terminals can perform relay functions when necessary, re-transmitting the signal from the base station to the other mobile terminals.

This makes it possible to guarantee a large coverage radius with a very low transmitted power level, as each relay operates as if it were a tiny base station.

Common channels

The following common channels are provided:

- BCH (Broadcast Channel) – used on the down-link to transmit system information in the entire cell.

- FACH (Forward Access Channel) – used on the down-link to transmit control information to a mobile terminal in cases where the system knows the cell in which the terminal is registered. It can be used to transport short data packets (as with the GSM Short Message Service).

- PCH (Paging Channel) – used on the down-link to transmit control information to a mobile terminal whose location is not known. Transmission is associated with a physical channel, the paging indicator, which informs the mobile terminal to which the message is addressed that information is present on the paging channel, thus permitting lower battery consumption. Here again, short data packets can be associated with the transmitted message.

- SCH (Synchronisation Channel) – used on the down-link to permit synchronisation between the mobile terminal and the base station.

- RACH (Random Access Channel) – used on the up-link to transport control information transmitted by the mobile terminal (usually a request for access to the physical medium in order to begin a communication). Short data packets can be associated with the transmitted message.

- ORACH (ODMA Random Access Channel) – used in TDD mode to start a relay connection.

- CPCH (Common Packet Channel) – used on the up-link to transport data packets. It features contention access and is used to transmit burst traffic. It is always associated with a dedicated down-link channel on which the associated physical layer signal-

Close-up 3.3 – Packet service transmission

In the multimedia age, it is likely that packet services will continue to grow in importance. As a result, a number of methods for transmitting packet services have been specified.

Three different transmission modes are envisaged, depending on the size of the packets to be sent and their arrival rate.For short, infrequent packets, data can be transmitted on common signalling channels (PCH, FACH and RACH), as it is now currently done for the GSM SMS service. For infrequent packets whose size can call for high bit rates, the Downlink Shared Channel (DSCH) and the contention up-link channels (CPCH and USCH) are used. An interesting feature of the DSCH is that it is always associated with a dedicated channel. The latter is used to transmit the physical layer signalling associated with the packets to be transmitted, as well as to provide services with real time constraints. In the case of an audio–visual service, for example, the video signal can be transmitted on the DSCH, while the audio signal can be transmitted on the associated DCH channel, thus guaranteeing the delay constraints. Likewise, two independent communications to the same terminal can be provided (thus making it possible to receive a voice call in the course of a data connection), separating the two data streams on the two channels.The third mode envisaged for transmitting packet services consists of setting up a dedicated channel. The packets are transmitted on this channel with the bit rate required by the service concerned. Instead of releasing the communication when packet transmission ends, however, the DPCCH channel is maintained in order to ensure synchronisation between the transmitter and the receiver, though the level of interference generated – and thus resource occupation – is reduced. The transmission resource is then re-engaged at the moment a new packet is transmitted. Conversely, if no packets arrive within a predetermined interval, the connection is released.

ling (power control commands) is transmitted. A channel with similar characteristics called the Uplink Shared Channel (USCH) is present in TDD mode.

- DSCH (Downlink Shared Channel) – used on the down-link to transport data packets. Access is shared by various users and is regulated by the base station.

Note that there are certain physical channels which are not associated with a transport channel. They are used to transport physical layer information that does not need to be sent to the higher layers. These channels are as follows:

- CPICH (Common Pilot Channel) – a down-link channel on which a known unmodulated sequence is transmitted. It makes propagation channel estimation possible.

- DPCCH (Dedicated Physical Control Channel) – physical channel present on both links and used to transport physical layer signalling.

- AICH (Acquisition Indication Channel) – present on the down-link and used to inform the mobile terminal that there is a message on the FACH channel in response to an access attempt.

- PICH (Paging Indication Channel) – present on the down-link and used to inform the mobile terminal that there is a message on the PCH channel.

The correspondence between transport channels and physical channels is summarised in Table 3.2.

3.3.2 Physical channels

The physical channels are typically based on the following structure:

- Radio frame: has a length of 10 ms and consists of fifteen time slots.

Table 3.2 Main UTRA parameters

Transport channel	Corresponding physical channel
BCH	*Primary Common Control Physical Channel (Primary CCPCH)*
	Common Pilot Channel, (CPICH) – present only in the FDD component
SCH	*Physical Synchronisation Channel (PSCH)*
FACH	*Secondary Common Control Physical Channel (Secondary CCPCH)*
PCH	Secondary *CCPCH*
RACH	*Physical Random Access Channel (PRACH)*
ORACH[a]	Present only in the FDD component and transmitted on the *PRACH*
CPCH (FDD)	*Physical Common Packet Channel (PCPCH)*
USCH (TDD)	*Physical Uplink Shared Channel (PUSCH)*
DSCH	*Physical Downlink Shared Channel (PDSCH)*
DCH	*Dedicated Physical Data Channel (DPDCH)*
ODCH[a]	ODMA – *Dedicated Transport Channel.* Present only in the TDD and transmitted on the DPDCH
	Dedicated Physical Control Channel (DPCCH). Present only in the FDD component[b]
	Acquisition Indication Channel (AICH). Present only in the FDD component
	Paging Indication Channel (PICH)

[a] This channel is present only in networks where the mobile terminals can also operate as relays to extend cellular coverage.

[b] In the TDD component, this information is transmitted in the DPDCH.

- Time slot – has a length of 10/15 ms. Each slot consists of a number of symbols which varies according to the bit rate of the service to be transmitted.

- Symbol – this is the information element after the channel encoding operations (i.e. after the error correction codes are inserted). Each symbol is multiplied by a number of chips equal to the spreading factor of the service to be transmitted in order to obtain a constant 2560 chips per slot.

Note that in the TDD component, the terms 'radio frame' and 'time slot' retain the same meaning assigned to them in the GSM system. However, their meaning differs for the FDD component. In this case, in fact, the various physical channels are transmitted on all the slots in the frame (one of the characteristics of the CDMA technique is that signals are transmitted continuously). The frame then takes on the meaning of the minimum transmission element in which the information transfer rate is maintained constant: the source rate can change at every frame (i.e. every 10 ms). Note, however, that the chip rate remains constant as the source rate varies. Similarly, the time slot is the minimum element of the physical channel for which the transmission power is maintained constant. By means of the power control mechanism, the transmission power can be increased or reduced at each slot.

Figure 3.5 shows an example of how the source bit rate can be varied on a radio at a rate equal to one frame basis (10 ms). The effects of power control on the power level are illustrated in Figure 3.6.

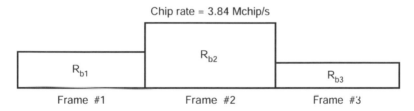

Figure 3.5 Source bit rate variation on a at a rate equal to one radio frame basis (10 ms).

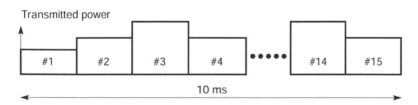

Figure 3.6 Variation in the power transmitted on a time slot basis as a result of power control.

Physical layer signalling

In the FDD component, physical layer signalling associated with the dedicated channels (DPDCH) is transmitted on the DPCCH channel. Physical layer signalling is transmitted in each slot. The following data fields are envisaged:

- Dedicated pilot symbols – transmitted in order to estimate channel impulse response. These symbols are also used to perform estimates of the interference level in order to establish the commands to be sent for power control.

- Transport Format Combination Indicator (TFCI) – these bits are used to indicate the type of service and the associated coding for the information transmitted on the frame. This field has been newly introduced for multimedia services. In order to vary the source rate rapidly while guaranteeing a larger number of formats at the same time, in fact, the receiver must be informed of the format of the transmitted data.

- Transmit Power Control (TPC) – in communications between two entities (A and B), these bits transport the power control commands. Entity A estimates the quality of the signal received by B and sends TPC commands so that B can vary its transmission power in order to reach the predetermined quality.

- FeedBack Information (FBI) – in communications between two entities (A and B), this field is used to transport information about the status of the signal received by B so that feedback loop can be closed and the format of the signal transmitted by A varied accordingly.

Other physical layer signalling (common pilot channel, PICH, AICH) is used to enable common channels to be received. For the secondary CCPCH channels, physical layer signalling consists of the pilot symbols and the TFCI field (given that different rates are envisaged); the TPC field is not needed, as power control is not provided for these channels.

In the TDD component, physical layer signalling is multiplexed with the DPDCH channel, since explicit DPCCH channel transmission is not envisaged. Signalling consists of a known sequence (the midamble) transmitted in the centre of the time slot and of the TFCI and TPC fields. The midamble is used to estimate the propagation channel, while the other fields have the same function used in the FDD component.

Spreading sequence assignment

In order to transmit signals with a variable source rate, it is necessary to use spreading sequences with variable length (the chip rate, in fact, is constant at 3.84 Mchip/s).

The range over which sequence length can vary differs for the FDD and TDD modes. For FDD, length can vary from a minimum of 4 to a maximum of 512, while for TDD it varies from 1 to 16.

To reduce the level of interference between users, each transmitted signal must be associated with a sequence, which is orthogonal to the others that have already been assigned (i.e. with zero cross-correlation).

This can be accomplished by adopting a sequence assignment based on a tree structure, where the sequences located on the branches are generated from a common root (see Figure 3.7). The sequences of equal length and those which differ in length but are generated by different roots of the tree are orthogonal to each other. As the number of available sequences limits the number of simultaneously transmitted signals, it is necessary to guarantee that the sequences are carefully managed. In general, sequence allocation to the various signals must be reviewed with a certain frequency rate in order to optimise resource utilisation and thus capacity.

3.3.3 Transmission of multimedia services with different quality requirements

In order to guarantee transmission of multimedia services, methods for simultaneously transmitting dedicated channels which call for different quality requirements have been developed.

In general, the quality requirements associated with the physical layer are as follows:

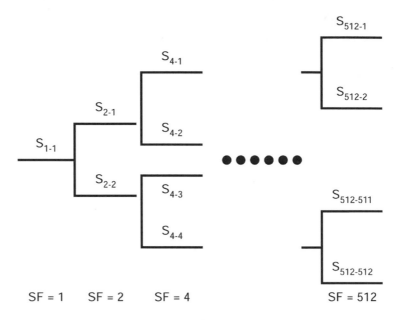

Figure 3.7 Spreading sequence tree.

- Error rate – the maximum error rate that can be supported to guarantee quality of service.

- End-to-end delay – the maximum delay that can be supported by the service in question.

These parameters have an impact on the type of coding used to correct the errors introduced by the propagation channel, on the depth of interleaving blocks, and on the maximum permissible number of repeats in cases where ARQ (Automatic Repeat reQuest) techniques are employed.

Simultaneous transmission of dedicated channels with different quality requirements is based on the scheme shown in Figure 3.8. The different services are encoded and interleaved independently on the basis of their quality requirements (note that the greater the interleaving depth, the better performance will be in terms of error rate in a mobile radio channel, though the amount of delay introduced will

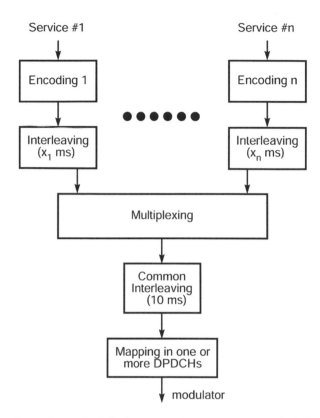

Figure 3.8 Operating principle for transmitting services with different quality requirements.

be increased). At this point, the different services are multiplexed, and the stream thus obtained is then interleaved (with depth equal to one radio frame) and mapped onto one or more DPDCH channels, depending on the rate achieved. A single code is used for symbol rates which call for a spreading factor of not less than 4. For higher rates, several spreading sequences are assigned to the same signal, thus generating more than one DPDCH channel.

3.3.4 The modulator

In the FDD component, two different modulation schemes are used

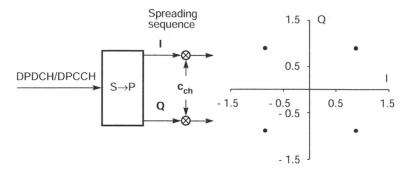

Figure 3.9 Down-link modulator and transmitted signal constellation.

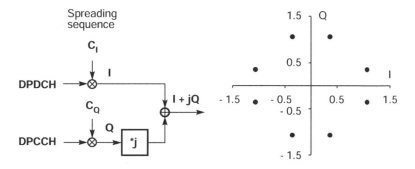

Figure 3.10 Up-link modulator and transmitted signal constellation.

for the up-link and down-link. Specifically, the down-link uses classic QPSK (Quadrature Phase Shift Keying) modulation (see Figure 3.9). The data stream obtained by time-division multiplexing the DPDCH and DPCCH channels is divided between the two components of the modulator (in-phase, I, and quadrature, Q). The resulting two streams are then multiplied by the spreading sequence and transmitted.

On the up-link, the approach used is called dual code BPSK (Binary PSK). In this case, the DPDCH channel and the DPCCH channel are transmitted independently on modulator component I and on component Q (see Figure 3.10). The two data streams are multiplied by two different spreading sequences (C_I and C_Q) with, in general, a different spreading factor. The spreading factor for the DPCCH channel is always 256, while that of the DPDCH channel varies in accordance with the data rate. The DPCCH is then multiplied by a gain factor (of less than one) in order to reduce the interference generated on the DPDCH channel and transmitted signal envelope fluctuations. Figure 3.10 illustrates the constellation obtained when the DPCCH channel is transmitted with a power 3 dB below the DPDCH.

As the fundamental difference between the two links, the DPCCH channel is transmitted in different ways: on the down-link, it is multiplexed with the DPDCH using time-division techniques, whereas on the up-link it is transmitted as if it were an independent channel. This difference is associated with the fact that in the event of discontinuous transmission (i.e. a source bit rate of zero), the square wave thus generated on the up-link would cause audio band interference, as occurs with GSM. With this solution, the transmitted signal is never interrupted, so there are fewer electromagnetic compatibility problems.

The TDD component uses the QPSK technique on both the up-link and the down-link.

3.3.5 The receiver

The receiver is not described in specification documents, as it is up to the manufacturer to optimise its performance. What must be specified, however, are the procedures for performing the measurements envisaged for the receiver. Of these, the most important are the measurements associated with hand-over and power control. Though the receiver is not specified, two types of structure are commonly found in the literature: the Rake receiver and reception techniques with interference cancellation.

In mobile radio systems, the radio channel propagates on multiple paths as a result of electromagnetic wave reflection and diffraction on obstacles such as buildings, trees and so forth. In practical terms, assuming that a single pulse is transmitted, what is detected at the receiver is a sequence of pulses whose amplitude, phase and delay vary over time according to precise statistical laws. The effect on the radio channel is illustrated in Figure 3.11, where α_i, τ_i and φ_i are, respectively, the amplitude, delay and phase of the i-th path.

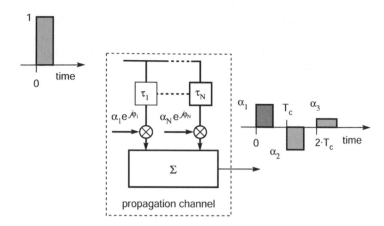

Figure 3.11   Effect of multipath propagation on the transmitted pulse.

The dispersion in transmitted energy over time translates into inter-symbol interference: in other words, a transmitted symbol interferes with a certain number of symbols that follow it. In narrow-band systems such as GSM, equalisation is used to counteract inter-symbol interference. The equaliser estimates the propagation channel and attempts to cancel symbol overlap.

In wide-band systems such as CDMA, transmitted energy dispersion over time can be an advantage, as the properties of the spreading sequence can be used to separate the various echoes and recombine their energy constructively. In practice, the receiver's resolution is in the order of one chip period (the inverse of the chip rate).

As the amplitudes of the various echoes are statistically independent, constructive recombination increases the probability that the signal will reach the receiver with sufficient amplitude for correct reception. This property is called 'path diversity', and can be used to advantage by means of the Rake receiver.

Essentially, the Rake receiver consists of many independent receivers, each tuned to a different replica of the signal. Like the garden tool of the same name, the Rake receiver aligns the different contributions and collects the energy from the multiple propagation paths on the teeth of the rake, or fingers as they are called in this case.

After the despreading operation, a signal with a rate equal to the information symbol period is obtained for each rake finger. The phase shift introduced by the channel is then recovered on each finger, e.g. by means of a reference signal of known phase. Finally, the recombination device performs a weighted summing operation on the output signals from each finger and supplies a single value for the information symbol on which the threshold decision is to be made.

A block diagram of a Rake receiver is shown in Figure 3.12, where T_c is the chip period.

The other form of receiver found in the literature is based on reception techniques with interference cancellation. The capacity and performance of a CDMA system are generally limited by the level of interference generated by the other users.

The Rake receiver, which extracts only one the information-bearing signal from the sum of all received signals and considers all the others signals as interference. The interference cancellation and joint detection techniques consider all signals to be information-bearing. To carry out this operation, the knowledge of the spreading sequences used to distinguish between among the various signals is required. As this knowledge is generally available in a CDMA system, it can be

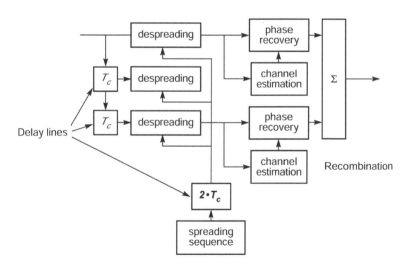

Figure 3.12   Schematic block diagram of a three-finger Rake receiver (T_c = chip period).

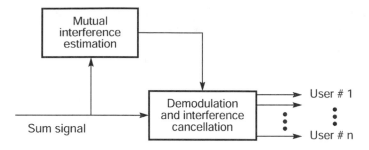

Figure 3.13   Functional diagram of a receiver based on joint detection.

used to demodulate all users of the same cell simultaneously. The functional diagram of a receiver based on joint detection is shown in Figure 3.13.

The major drawback of joint detection lies in the receiver's greater complexity. This complexity increases exponentially with the number of users to be demodulated simultaneously, and thus with the length of the spreading sequences. Consequently, it is preferable to apply this technique to systems with low spreading factors, as in the TDD component. One of the technique's main advantages is that, by cancelling interference, it makes the receiver less sensitive to the near–far effect and thus makes it possible to relax power control speed rate and accuracy requirements. In addition, better performance in terms of error rate translates into a proportional increase in system capacity.

3.3.6 Power control

Three power control procedures are envisaged for the FDD component: open loop power control, closed loop power control, and outer loop power control.

Open loop power control is used only on the up-link, when the call is being set up. The power with which the PRACH channel is transmitted is calculated on the basis of the power (known beforehand or sent over the broadcast channel) received on a predetermined common control channel transmitted by the base station. This procedure estimates down-link fading attenuation, which as an initial approximation is assumed to be equal to that on the up-link. The power level to be transmitted on the basis of open loop control is approximate. In fact, the frequency spacing between the carriers used for the two links is such that there is little correlation between the propagation phenomena on the up-link and on the down-link. Only over the long term fading on the two links can be assumed to be approximately equal, obviously taking the difference due to frequency spacing into account.

As the open loop procedure is not sufficient to guarantee the necessary accuracy, a more sophisticated solution must be introduced. The solution which has been adopted for this purpose is closed loop power control (see Figure 3.14) based on frequent transmission of commands to increase or decrease the power to be transmitted on the controlled link.

The frequency with which these commands are transmitted is equal to the time slot (i.e. a command is transmitted every time slot). The procedure measures the carrier-to-interferer ratio (C/I) on the controlled link and compares it with a reference value $(C/I)_{th}$. Depending on the outcome of this comparison, commands to increase or decrease power by a certain predetermined value (typical values range from 0.5 to 1.5 dB) will be sent on an appropriate control channel (DPCCH) for the opposite link.

Outer loop power control calculates the reference threshold $(C/I)_{th}$. This calculation is performed much less frequently than is needed for closed loop control. The threshold is updated on the basis of connection quality, which is monitored continuously. In general, changes are due to variations in the environment and to resource management actions as an ASU (Active Set Update).[1]

However accurate it may be, the power control procedure is never ideal. In practice, what is obtained is not a constant level of received power (and hence a constant E_b/I_0 ratio), but a value with a certain spread around the mean. An example is shown in Figure 3.15, which illustrates the probability density for the power transmitted by a mobile terminal and received by the radio base station. As can also

(1) See hand-over procedures.

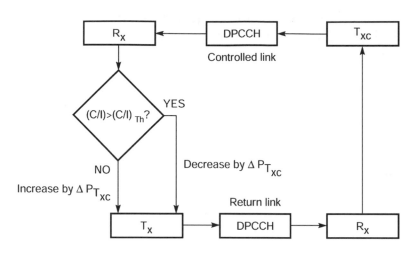

Figure 3.14   Operating principle of the closed loop power control procedure.

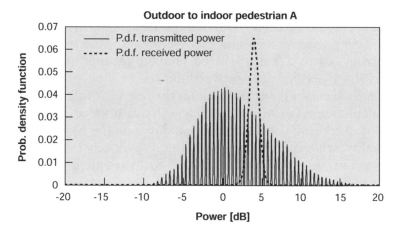

Figure 3.15   Probability density for transmitted power and received power with power control.

be seen from the figure, the transmitted power can variate within a wide dynamic range in order to compensate for the variations in the propagation channel and thus maintain the received power as constant as possible.

The performance of closed loop power control is affected by the mobile terminal's speed. In fact, the feedback introduces delays because of the need to perform channel measurements, to process them and to transmit generated commands (i.e. TPC bits on the DPCCH channel). One or two time slots may go by between the instant a power control command is generated and the one it is executed.

At low mobile terminal speeds, variations in propagation conditions are relatively slow, and can be accurately compensated by frequent power control commands. As the mobile terminal's speed increases, however, variations in propagation conditions take place at increasing frequency. Consequently, the delays introduced by power control make propagation channel tracking less and less effective. At high speeds (over 100 km/h), received power control commands begin to lose their correlation with the channel condition. As speed increases further, the commands are no longer correlated with the channel's actual condition, and the procedure thus leads to degraded performance. Theoretically, it would be better to disable the power control at this point. However, the required quality of service can be guaranteed through other mechanisms such as error correction codes and interleaving.

The fact that power control is not ideal also has an impact on cellular coverage. In a CDMA system, all cells generally share the same carrier. In practice, imperfect power control and other factors make it difficult to locate cells with highly dissimilar dimensions (i.e. in which the average transmitted power level differs widely) near to each other. Consequently, different carrier frequencies must be assigned to the three typical cellular coverage levels: picocell, microcell and macrocell. Say, for example, that we are in a suburban area where coverage passes from the microcell (urban) to the macrocell (rural) level. In general, while an imperfection in the power transmitted by a macrocell user has very little impact on the macrocell, the same cannot be said of any microcells bordering it. The fact that the microcell is geographically smaller than the macrocell means that an error originating in the latter can cause outages in the microcell: hence the need to assign different carriers to different hierarchical levels.

In the TDD component, the requirements for power control are less stringent. Thanks to the TDMA technique, in fact, signals transmitted on different time slots do not interfere with each other. Nevertheless, the advantages of power control are such that the procedure is also applied in to this component, using the three methods described above.

In this case, given that the signal is transmitted and received at the same frequency, open loop power control can be used more successfully. Closed loop power control, by contrast, is less effective: assuming that transmission takes place on a single time slot per frame, commands can be sent at a minimum rate of one every 10 ms. Consequently, open loop control is also applied to the dedicated traffic channels (DCH).

References

[1] 3G TS 25.201, 3rd Generation Partnership Project; Technical Specification Group Radio Access Network; Physical layer – General description; (Release 1999).
[2] 3G TS 25.211, 3rd Generation Partnership Project; Technical Specification Group Radio Access Network; Physical channels and mapping of transport channels onto physical channels (FDD); (Release 1999).
[3] 3G TS 25.221, 3rd Generation Partnership Project; Technical Specification Group Radio Access Network; Physical channels and mapping of transport channels onto physical channels (TDD); (Release 1999).
[4] 3G TS 25.213, 3rd Generation Partnership Project; Technical Specification Group Radio Access Network; Spreading and modulation (FDD); (Release 1999).
[5] 3G TS 25.223, 3rd Generation Partnership Project; Technical Specification Group Radio Access Network; Spreading and modulation (TDD); (Release 1999).
[6] 3G TS 25.990, 3rd Generation Partnership Project; Technical Specification Group Radio Access Network; Vocabulary; (version 3.0.0).

[1–6] © ETSI 1999 Further use, modification, redistribution is strictly prohibited. The above mentioned standard may be obtained from ETSI Publication Office, publication@etsi.fr, Tel.: +33 (0) 4 92944241.

4

The UMTS Access Network

Sergio Barberis, Daniele Franceschini,
Nicola Pio Magnani and *Enrico Scarrone*

4.1 Introduction

Like other cellular mobile radio systems, UMTS is often identified
with its radio features, forgetting that there is also a large and
complex network transporting the bulk of the information, be it
voice or data, coming from the mobile terminals.

The *Core Network* is the component of this network that establishes
communication between the various sections of the *Access Network*,
which gathers traffic directly from the various radio base stations.

As was indicated in the previous chapters, the UMTS Terrestrial
Radio Access Network is called UTRAN.

The UTRAN role is depicted in Figure 4.1, which illustrates the
UMTS high-level architecture as a function of the following logical
entities: UE (*User Equipment*, or in other words the mobile terminal),
UTRAN and CN (*Core Network*). In UMTS, moreover, an attempt has

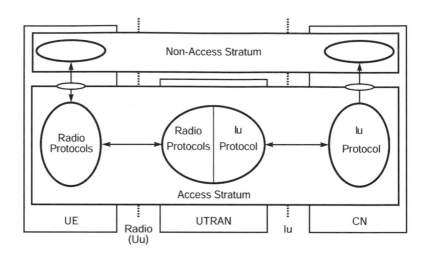

Figure 4.1   High-level UMTS functional architecture.

been made to introduce a certain degree of independence between the radio interface and the other parts of the system. This independence, investigated in the framework of specific research programs, has been partially implemented by means of logical separation between the UMTS *Access Stratum* and *Non Access Stratum*. Here, the *Access Stratum* is the set of protocols and capabilities which are most closely linked to the considered radio technique, while the term *Non Access Stratum* is used to denote those which are independent of the radio access network.

Theoretically, this permits a mobile radio system to use different access networks, freeing the Core Network from the particular technology chosen for the access. As a result, several types of access network can be connected to the same mobile radio system.

The Non Access Stratum capabilities include call and session control (i.e. the procedures used to set up, modify and release the transmission logic resources relevant for the required service) and mobility control (or in other words, all the procedures which enable the user to communicate regardless of his location and whether or not he is in motion). This latter capability applies to mobility between the various areas of the access network, and as such is managed by the CN. By contrast, mobility within an access area is managed independently in the Access Stratum. Essentially, the Access Stratum capabilities correspond to the set of functions implemented in the UTRAN.

To provide an overview of the radio access network and highlight its innovative aspects, the following topics will be investigated in depth below, based on 3GPP Specifications [1–3]:

- General architecture of UTRAN, with a detailed look at AAL2 (*ATM Adaptation Layer 2*) and macrodiversity.

- Architecture of the UTRAN protocols.

- Architecture of the radio protocols, with particular attention to the RRC (*Radio Resource Control*), RLC (*Radio Link Control*) and MAC (*Media Access Control*) layers and the interactions between them.

4.2 UTRAN architecture

As can be seen from Figure 4.1, the radio access network is bounded by two interfaces: on one side, the U_u radio interface stands between UTRAN and the mobile terminals, while on the other side, the I_u interface connects UTRAN to the Core Network. Actually, the latter interface performs a dual function, as it integrates both the interface that connects UTRAN to the circuit based CN (CS – *Circuit Service*) and the interface connecting UTRAN to the packet based Core Network (PS – *Packet Service*). Figure 4.2 illustrates the UTRAN structure in detail.

As we have seen, UTRAN consists of a set of *Radio Network Subsystems* (RNSs) connected to the CN via the I_u interface. An RNS consists of a controller (the *Radio Network Controller*, or RNC) and one or more entities called *Nodes B*, which are connected to the RNC through the I_{ub} interface. A Node B superintends a set of cells which may be FDD (*Frequency Division Duplex*), TDD (*Time Division Duplex*), or mixed.

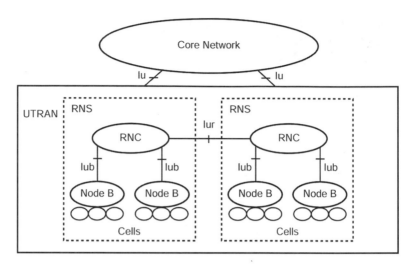

Figure 4.2   UTRAN components and interfaces.

In UTRAN, different RNCs can be connected to each other through the I_{ur} interface.

It should be kept in mind that the RNC is the boundary between the radio domain and the rest of the network. The protocols opened in the terminal to manage the air link (i.e. the radio protocols which cross the I_{ub} and I_{ur} interfaces) are terminated in the RNC. Above the RNC are the protocols that permit interconnection with the CN and which are depending on it. It should be clarified that the wireless link is limited to the terminal-base station link; thus, the radio protocols are referred to by this name, not because the associated messages travel on the radio link, but because they manage the radio interface.

In addition to permitting scalable RNS sizing, this architecture provides several other advantages, including a significant capacity for managing mobility inside UTRAN (and hence in the Access Stratum level).

In fact, both Node B and the RNC are capable of managing hand-over and macrodiversity. Hand-over is the capability of mobile radio systems that enables a radio connection to be maintained when the user moves from one cell to another. Macrodiversity refers to the ability to maintain an ongoing connection between the mobile terminal and the network though more than one base station; this capability is particularly important in CDMA systems. The hand-over and macrodiversity functions and the advantages that derive from their interaction are discussed in greater depth at the end of this section.

Hand-over and macrodiversity can be managed at Node B level (for cells belonging to the same Node B), or can be managed at RNC level by using the I_{ub} interface (for cells that belong to different Nodes B but are controlled by the same RNC) or the I_{ur} interface (for cells belonging to different RNSs).

Between different RNSs, hand-over can also be accomplished via the CN by using the I_u interface. In this case, however, macrodiversity is not possible, as it is implemented by means of the radio protocols that are limited to the RNC. This point will be clarified below.

In effect, the real reasons for the existence of the I_{ur} interface are associated with mobility management within UTRAN. However large each RNS may be in terms of geographical territory and the number of users served, it cannot satisfy all mobility needs. The I_{ur} interface on the one hand permits continuous mobility, with transitions between RNSs that are not perceptible to the user (thanks to macrodiversity); on the other hand, it lightens the burden on the CN procedures, limiting the cases in which the CN must take action to those in which this interface is not present.

Another feature which distinguishes UTRAN is the choice of transport protocols on the I_u, I_{ub} and I_{ur} interfaces. The protocols are based essentially on ATM, with information streams adapted to ATM characteristics using the *ATM Adaptation Layer 2* (AAL2) to transport the radio protocols (I_{ub} and I_{ur}) and user streams to the Circuit Service (I_u), and using IP over AAL5 for user streams to the Packet Service (I_u).

The following Close-ups focus on how AAL2 is used by UTRAN, and on the concepts of soft hand-over and macrodiversity in UTRAN. The latter will receive special attention, as they are among the UMTS salient features.

Close-up 4.1 – AAL2 in UTRAN

In UTRAN, it was decided to adopt the ATM transport technique in order to have a flexible mechanism that can potentially be adapted to different combinations of multimedia traffic. However, it was necessary to make allowance for the particular characteristics of traffic components such as voice, which are characterised by a low bit rate and are subject to real time requirements.

If, in fact, we consider the compressed voice used in the world of mobile radio, and assume an 8 kbit/s bit rate with a 10 ms frame length, it is likely that an 80 bit voice packet must be sent every 10 ms. Using ATM cells with conventional AAL is clearly wasteful: the ATM cell has a payload of 48 bytes (384 bit), so that only slightly over one fifth of capacity is used (as will be recalled, an ATM cell consists of a header with control and packet identification information, and a component containing the data which is effectively transported, i.e. the payload).

AAL2, however, can multiplex different users' traffic on the same cell stream, both by putting packets from different users in the same cell, and by dividing a user packet among two cells as illustrated in Figure 4.3.

Close-up 4.2 – Soft hand-over and macrodiversity

Some of the concepts introduced in Chapter 2 will be reviewed below. In a cellular mobile radio system using the CDMA access technique, all cells (or at least those on the same hierarchical level) operate on the same carrier. This means that the signal transmitted by a radio base station can be received and decoded by anyone who knows the spreading code used and is located in the geographical area in which the signals are received above a certain minimum power threshold. The concept of macrodiversity is thus a natural addition to this type of system: to improve communication quality, the mobile terminal does not limit itself to remaining connected to a

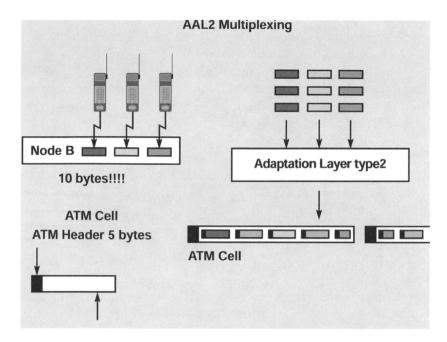

Figure 4.3   Example of AAL2 multiplexing.

single BS (*Base Station*) as usually occurs in TDMA systems, for example, but involves all BSs from which it receives a sufficiently good reference signal in the communication. We thus have the concept of the *Active Set* (AS), which is defined as the group of BSs from which the mobile terminal receives the same useful information. The process of activating and releasing parallel links is carried out as a function of the quality of the signal received by the mobile terminal as it moves in the network. Note that the procedures with which the connection is routed in the network depend on the mechanisms used to form the active set and on the network topology. When the mobile terminal operates with macro-diversity, it can improve communication quality on the down-link by combining the signals with highest energy content. These signals can come from cells belonging to different Nodes B and RNCs. They are handled in such a way as to be 'time aligned' upon reaching the mobile terminal. In this way, several signals with the same information content can be combined.

On the up-link, or in other words between the mobile terminal and the base station, the signal transmitted by the mobile terminal is decoded by all of the BSs in the active set, so that the signal is recombined at the next higher hierarchical level (Node B, if the BSs in the AS belong to the same Node B, or RNC, if the BSs in the AS belong to different Nodes B). This makes it possible to obtain a significant improvement in quality. The amount of improvement also depends on the type of signal recombination: it is possible to use a simple selection mechanism based on picking the signal containing fewer errors (macrodiversity implemented at RNC level), or all of the received signals can be used, combining the different contributions (macrodiversity implemented at Node B level).

Using macrodiversity makes the hand-over operation particularly simple, which is why we speak of *soft hand-over* in CDMA systems. It is also why the terms macrodiversity and soft hand-over are often confused in current usage. Further information about soft hand-over is provided in Section 4.4.3.

From another standpoint, using macrodiversity in the network makes it possible to increase system capacity. In fact, macrodiversity can also be turned to advantage in guaranteeing that the desired quality target can be reached even when the power transmitted by the mobile terminal is reduced, using the combination of signals from the AS base stations. In this way, the system operates at lower power levels than it would if macrodiversity were not used. As CDMA is a power limited system, this results in an increase in the system overall capacity.

Another of the advantages provided by macrodiversity is the system greater robustness against shadowing, i.e. the signal attenuation as a result of obstacles between the mobile terminal and the BS. It is self-evident, in fact, that if the mobile terminal is connected to several BSs at the same time, it is much less likely that there will be obstacles on all of the links between it and the base stations.

4.3 UTRAN protocol architecture

The architecture of the UTRAN protocols is shown in Figure 4.4.

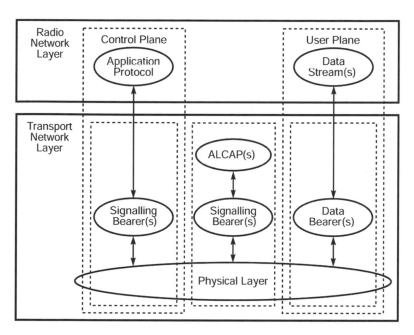

Figure 4.4   UTRAN interface protocol architecture.

The first clear division is between the control plane and the user plane. The former carries signalling (control) information, while the latter carries user information. A further division is between the *Radio Network Layer* and the *Transport Network Layer*. For UTRAN, the latter essentially provides the next higher layer with two types of transport channels, or bearers.

The first of these types is dedicated to data transport, and it is thus called the data bearer. It transports radio protocol streams on the I_{ur} and I_{ub} interfaces (AAL2/ATM), as well as circuit switched (AAL2/ATM) and packet switched (IP/AAL5/ATM) user data streams on the I_u interface.

The second type, or signalling bearer, is based on SS#7/AAL2/ATM (or on IP/AAL5/ATM). Signalling concerns the network application protocols which establish, re-establish (during hand-over) and release the bearers whenever required, and manage all of the associated mobility procedures.

To complete the information shown in the figure, it should be emphasised that these channels can either be established beforehand through management procedures, or set up on request via signalling procedures. In the latter case, a signalling protocol defined generically as ALCAP (*Access Link Control Application Protocol*) is used. In reality, this is identified with the AAL2 signalling protocol.

The point that should be stressed here is that the radio protocols (described in paragraph 4.4) are interpreted from the standpoint of the I_{ur} and I_{ub} interfaces as data streams to be transported. As a result, their control and transport components are handled in the same way. The situation differs on the I_u interface, which, being located above the radio protocol termination point (the RNC), handles control information separately from the user data transport streams.

We will not go into details of these protocols and of the I_u interface here. It is sufficient to say that a different application protocol is specified for each interface as follows:

- RANAP (Radio Access Network Application Part) on the I_u interface.

- RNSAP (Radio Network System Application Part) on the I_{ur} interface.

- NBAP (*Node B Application Protocol*) on the I_{ub} interface.

More detailed information about these protocols can be found in the 3GPP UTRAN Specifications.

4.4 The radio protocols

After describing the architecture of the UTRAN protocols, we will now turn to the various radio access protocol layers. The radio protocols (which terminate well beyond the wireless portion, going as far as the RNC) implemented in the UMTS network nodes are specified on several layers. In the more complex layers, there is a further division in sub-layers which corresponds to a precise functional division. In particular, the following sections will start from a general description of the radio interface protocol structure and proceed with a detailed discussion of the interactions between the various protocol layers, the issues involved in managing radio resources, and the capabilities and services provided by each layer in the protocol, highlighting their actual role in UTRAN.

4.4.1 Radio protocol architecture

Figure 4.5 provides an overview of UTRAN radio protocol architecture. The radio protocols can be broken down into three layers: the Physical Layer or Layer 1, the Data Link Layer or Layer 2, and the Network Layer or Layer 3.

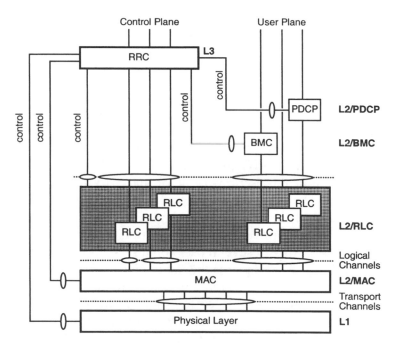

Figure 4.5   UTRAN radio protocol architecture.

Figure 4.5 also shows how the Access Stratum protocols are divided into two planes: the Control Plane and the User Plane. The information carried by the access network may be signalling (control) information or user information, i.e., deriving from the user service. The purpose of signalling is to create the conditions for correct user information transmission. Layer 3 is basically responsible for managing signalling in UTRAN, and it is thus located in the Control Plane. By contrast, layers 1 and 2 also perform transmission functions and thus provide a transport base for both signalling and user information. In this sense, layers 1 and 2, cut across the Cotrol and User planes.

After the analysis of the physical layer main features provided in Chapter 3, we will now focus on the access network.

Layer 2 (the Data Link layer) is divided into two main sublayers: the MAC (*Medium Access Control*) sublayer and the RLC (*Radio Link Layer*) sublayer. Also PDCP (*Packet Data Convergence Protocols*) and BMC (*Broadcast/Multicast Control*) belong to layer 2.

The MAC is the layer which, in a telecommunications network, manages simultaneous access by a number of users (multiple access) to a shared resource. In the case of UTRAN, this shared resource consists of the radio resources, which are particularly precious because they are scarce. In this context, the MAC layer assumes an essential role which becomes even more significant if we consider the importance that the data transmission service has for UMTS. This type of service can be characterised by high bit rate transmission for short periods, with long inter-arrival times between the packets to be transmitted. Consequently, the idea is to use resources efficiently, employing them in the users' periods of inactivity to transmit data for other users. The situation we have just described highlights one of the typical tasks of the UMTS MAC layer: providing a radio access interface which is optimised for transmitting packet data through statistical multiplexing of several users on a set of common channels.

The other Layer 2 sub-layer is the *Radio Link Control* (RLC), which ensures that information is reliably transmitted in UTRAN and provides a retransmission service for those packets which Layer 1 was unable to deliver correctly to destination. The retransmission mechanism is a form of data protection that overlaps the one provided by channel coding at layer 1. In this way, for example, it is possible to guarantee data integrity under unfavourable mobile radio channel conditions. The *Radio Link Control* also performs an encryption function (in cases where this is not carried out by the MAC layer) in order to protect user information from unwanted interception.

Packet Data Convergence Protocol (PDCP) provides transmission/reception of network PDUs in acknowledged, unacknowledged and transparent mode. To do so, PDCP maps Network PDUs from a network protocol to a RLC instance. Also compression and decompression of not strictly network PDU control information (header compression/decompression) is performed here.

Broadcast/Multicast Control (BMC) provides a service for the broadcast/multicast transmission for common user data in transparent or unacknowledged mode.

From the standpoint of the radio protocols, layer 3 (the Network layer) contains the *Radio Resource Control* (RRC). The RRC functions play a fundamental part, as they are responsible for managing radio resources. In doing so, the Radio Resource Control is in charge of managing and controlling the lower MAC and RLC layers and the physical layer, as well as procuring the input parameters for the RRM

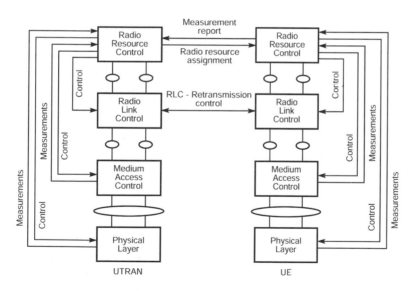

Figure 4.6   Interactions between UTRAN sub-layers.

(*Radio Resource Management*) algorithms and procedures. On the basis of the services requested by the higher layers and the actual allocation of the resources present in the cell at the time concerned, the Radio Resource Control assigns an RB (*Radio Bearer*) whose characteristics in terms of bit rate and QoS are such that it can support the required service. Once resources have been allocated, the Radio Resource Control uses the measurements received from the MAC and the physical layer to monitor resources so that they can be reconfigured dynamically and thus optimise their management.

4.4.2 Interactions between the radio protocol layers

Figure 4.6 depicts the interactions between the UTRAN radio protocols. In the model chosen for UTRAN, the RRC layer performs a true supervision of the other sub-layers, exerting full control over all of the UTRAN capabilities needed to allocate and manage radio resources. The RRC sub-layer implements the RRM manager, i.e. the set of all algorithms and procedures for interlayer radio resource management. In order to exercise full control over the other layers as shown in Figure 4.6, control connections are envisaged to the Radio Link

---

Close-up 4.3 – The RRC (Radio Resource Control) layer

The main functions provided by the Radio Resource Control layer are as follows.

Functions associated with mobility in an RRC connection: RRC evaluates, decides on and executes the procedures associated with mobility (such as hand-over, cell selection and reselection, which will be discussed in Section 4.4.3).

QoS (Quality of Service) request control: this function guarantees that the QoS required by the Radio Bearer needed to provide the required service is supported. This includes checking that enough radio resources are allocated.

Control over measurements performed by the mobile terminal: the measurements performed by the mobile terminal are under the full control RRC as regards which parameters are measured and how and when they are measured. In addition, the RRC layer also carries this information from the mobile terminal to the network.

Services provided by the Radio Resource Control

The Radio Resource Control provides three types of service to the higher layers:

- *General Control Service Access Point.* Broadcasts non-access stratum information to all users in a given geographical area.

- *Notification Service Access Point.* Provides *paging* and *notification.* Paging involves sending information inside a given area but which is addressed to a specific mobile terminal.

- *Dedicated Control Service Access Point.* Provides a service for setting up and releasing a connection and for transferring messages by means of this connection.

Close-up 4.4 – The RLC (Radio Link Control) layer

The RLC layer is the Layer 2 sub-layer which receives the services provided by the MAC and in turn provides services to the higher layers. Together with the ability to set up and release RLC layer connections on request, the Radio Link Control offers three different data transmission modes.

Transparent data transmission: this service transmits data from the higher layers without adding further protocol information of any kind.

Unacknowledged data transmission: this service transmits data from the higher layers without guaranteeing packet delivery to the corresponding RLC *Peer Entity.*

Acknowledged data transmission: this service transmits data from the higher layers, guaranteeing delivery of the transmitted information. If the receiving entity's RLC layer does not receive data correctly, it will notify the transmitting entity of the fact.

In addition to these transmission services, the RLC layer can also notify the higher layers of any errors which cannot be managed and eliminated by normal procedures. The RLC layer has a series of properties which are closely associated with the services it supports. The most important of these properties are as follows.

Segmentation and re-assembly: the RLC sub-layer supports transmission of variable length PDUs, or *Protocol Data Units*, from the higher layers. To adapt these PDUs to those of the RLC layer, the PDUs arriving from the higher layers are segmented and inserted in the corresponding RLC PDUs. In reception, the Radio Link Control performs the opposite operation, reassembling the information carried in several RLC PDUs in a single, higher layer PDU.

Concatenation: in the segmentation operation, if the highest-layer PDUs are not a whole multiple of the RLC PDUs, the first segment of the next PDU can be concatenated in the same RLC PDU with the last segment of the preceding PDU.

Padding: When concatenation cannot be applied, the remaining data to be transmitted may not be able to fill an entire RLC PDU. The rest of the PDU can then be filled with padding bits whose only purpose is to fill the part which would otherwise be empty.

Error correction: this function provides error correction by means of an appropriate repeat mechanism such as Selective Repeat, Go Back N, Stop And Wait ARQ.

Control, the MAC and the physical layer. In addition, the RRC layer needs measurements regarding resource use provided by the MAC and the physical layer for correct overall management of radio resources.

Apart from its formal position in layer 3 (the Network layer), radio resource management cannot be correctly located in any single layer of the OSI stack. Rather, this management should be regarded as an

Close-up 4.5 – The MAC (Medium Access Control) layer

The MAC layer is a sub-layer of the radio access network layer 2 (Data Link). It receives services from the physical layer by means of transport channels and provides services to the RLC sub-layer via logical channels as illustrated in Figure 4.5. While the transport channels indicate how information is transmitted, the logical channels reveal the kind of information that is transported.

MAC layer functions

The MAC layer performs an essential role in several processes: in regulating multiple user access to the same radio resources, in dynamically managing the radio resources as directed by the Radio Resource Control, and in implementing a series of procedures for a more effective resource management. Thus, the MAC layer provides the RLC layer with the services described below.

Data transmission: this service provides data transfer between MAC peer entities in unacknowledged mode. Data segmentation is not performed. Consequently, segmentation/re-assembly must be carried out by the higher layers.

Reallocation of radio resources and MAC layer parameters: on request from the Radio Resource Control, this service reallocates radio resources and changes MAC layer configuration parameters, such as the UE identity, the *Transport Format Combination Set*, and the type of transport channel.

Traffic and quality parameter measurements: the MAC layer performs a series of local measurements on traffic volumes and on the parameters that indicate transmission quality. Measurements are sent to the Radio Resource Control, which interprets the results of the received measurements and takes any necessary action, e.g. reconfiguring the underlying layers.

In order to effectively perform the tasks envisaged by the services described above, the MAC layer has the following capabilities.

Selection of the appropriate transport format for each transport channel on the basis of the instantaneous source rate: UTRAN can multiplex several DCH transport channels on the same physical channel. Multiplexing, which makes it possible to optimise the use of physical channels, can be regarded as a good example of interaction between layers. Each transport channel, in fact, is associated with a *transport format combination* consisting of a set of parameters providing information about the bit rate, the QoS and physical layer parameters. In order to multiplex several DCHs on the same physical channel, the channels must have homogeneous characteristics. For this reason, the Radio Resource Control provides the so-called *transport format combination set*, or in other words the set of *transport format combinations* for each DCH that makes multiplexing possible. The MAC layer has the task of selecting the appropriate transport format combination within the transport format combination set in relation to the source rate.

UE data stream priority management: the MAC layer can take advantage of this selection capability to assign different priorities to different streams from the same mobile terminal. Priorities can be determined on the basis of RB service attributes or the status of the Radio Link Control buffer.

UE priority management through dynamic scheduling: in order to make full use of spectral resources for data transmission, the MAC layer uses a dynamic scheduling method. The MAC layer manages priorities on common and shared channels by means of appropriate algorithms and procedures.

UE identification on the Common Transport Channel: when a given mobile terminal is addressed on a common down-link channel, or when the terminal accesses the RACH channel, it must be appropriately identified. Given that the MAC layer deals with access and multiplexing on the transport channels, it is natural for the identification function to be performed by the MAC layer.

Higher-layer PDU multiplexing/de-multiplexing on the Common Transport Channel: the MAC layer performs common transport channel multiplexing, since the physical layer does not support multiplexing for this type of channel.

Higher-layer PDU multiplexing/de-multiplexing on Dedicated Transport Channels: the MAC layer permits multiplexing of dedicated transport channels. This function can be useful when a considerable number of services from the higher layers can be mapped on the same transport channel. In these cases, the multiplexing identifier is contained in the MAC control protocol information.

Logical channels and mapping with transport channels

The MAC layer provides the RLC sub-layer with data transfer services on logical channels. To accommodate the different types of information transmission service offered by the MAC layer, a complete set of logical channels has been specified where each logical channel is distinguished on the basis of the transferred information. Logical channels can be grouped into two broad categories:

- *Control channels*, which transfer signalling information of the control plane.

- *Traffic channels*, which transfer information of the user plane.

The main control channels will be described below by way of example.

Broadcast Control Channel (BCCH): down-link channel which broadcasts system control information. Broadcast information could include: layer 3 information, system information and physical layer parameters required for network operation, e.g. down-link power level and up-link interference level.

Paging Control Channel (PCCH): down-link channel for the transfer of paging information. Used when the network is not aware of the mobile terminal cell location.

Common Control Channel (CCCH): bi-directional channel utilised for the transmission of control information between the network and the mobile terminal. It is used when an RRC connection has not yet been set up between the mobile terminal and the network.

Dedicated Control Channel (DCCH): bi-directional point-to-point

channel for transmitting control information dedicated to a particular user. It is set up by means of an RRC Connection Setup
procedure.

As an example, the description of the main traffic channel is given
below.

Dedicated Traffic Channel (DTCH): point-to-point connection
dedicated to a particular mobile terminal for transferring user
plane information. The DTCH channel exists on both the up-link
and the down-link.

Figure 4.7 shows an example of the criteria whereby the MAC
layer maps the transport channels provided by the physical layer
(see Chapter 3 for definitions of the latter channels) onto the main
logical channels described above. The standards for these criteria,
like those covering the number and functions of transport and logical channels, continue to be revised.

'interlayer' whose job is to co-ordinate the RLC, MAC and RRC sub-
layers.

Paragraph 4.4.3 will discuss mechanisms for dynamic resource
management through direct interaction among several layers. To

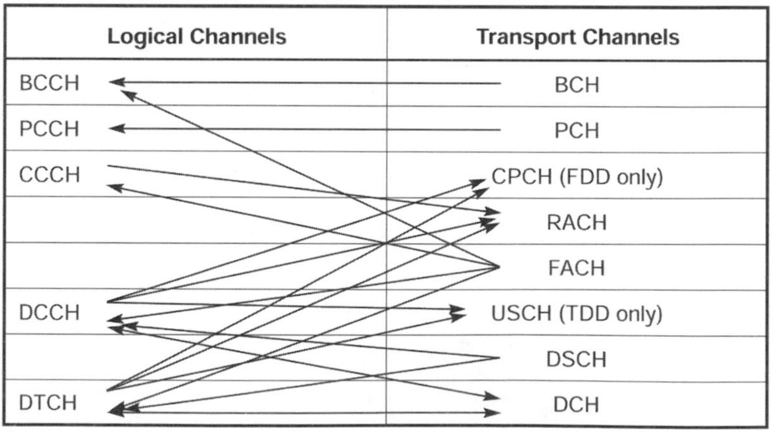

Logical Channels	Transport Channels
BCCH	BCH
PCCH	PCH
CCCH	CPCH (FDD only)
	RACH
	FACH
DCCH	USCH (TDD only)
	DSCH
DTCH	DCH

Figure 4.7 Example of correspondence between logical channels and transport
channels from the standpoint of the terminal.

provide a better grasp of the system workings, particular attention will be devoted to the services and capabilities of each of the sub-layers described.

4.4.3 Radio Resource Management (RRM)

Radio Resource Management (RRM) consists of the set of procedures and algorithms used for efficient management of the radio link layers. Implemented on the RRC layer, Radio Resource Management supervises and co-ordinates the functions provided on the other layers (MAC, RLC and physical layer), permitting correct and efficient use of the channels made available by the radio interface physical layer.

Radio resource management is a characteristic of all mobile radio systems. As was mentioned in the previous chapter, however, it is a topic of particular importance for the operation of a mobile radio

Close-up 4.6

To clarify the deep-seated differences between the way radio resources are managed in the UMTS, which uses the CDMA technique, and the way they are managed in a system based on FDMA/TDMA such as GSM, we will consider as an example the case in which the load on a system cell is particularly heavy. In such a situation, it would be advisable to relieve the cell overload by moving part of the traffic to the adjacent cells.

This can be accomplished fairly readily in a system like GSM by forcing some of the users to employ the resources of a radio base station other than the *best server* (i.e. the radio base station from which the mobile terminal receives the strongest signal). Overall, traffic will be distributed more uniformly in the system cells, with obvious advantages in terms of the total amount of traffic that can be handled. On the debit side, the mobile terminals which were forced to connect with a radio base station other than the *best server* will experience lower quality and will interfere to a greater extent (as they are closer in space) with those mobile terminals (and only those mobile terminals) which use the same radio frequency and time slot in other system cells. In particular, if a user of a system such as GSM asks to set up a call in a congested cell (or in other words, a cell in which all radio resources are already occupied), the system can attempt to satisfy the request by using the resources of another adjacent cell in what is called the *Direct Retry* procedure. Note that this attempt can be made even if the user in question is not near the cell border.

The same operation is far more delicate in a system based on the CDMA technique such as UMTS. In fact, the main objective of radio resource management in a CDMA system is to minimise the overall power transmitted by the mobile terminals (with reference to the up-link, which is affected by the near–far effect described in Section 3.1). Consequently, each user must be connected to the radio base station that makes it possible to use the lowest power. This radio base station is not necessarily the one from which the mobile terminal receives the strongest signal (i.e. the *best server* in GSM terminology). In fact, if the signal transmitted by the mobile terminal is to be correctly decoded, it must reach the radio base station with enough power to cope with the signals arriving from the other mobile terminals, which – by contrast with the GSM system – are not orthogonal in either the time domain or the frequency domain. The amount of power used by the mobile terminals is thus a function of the channel conditions and of the base station load level. As a result, each mobile terminal must be connected to the radio base station which makes it possible to minimise the sum (in logarithmic units) of the propagation loss and the level of interference, which is a function of load. This radio base station will be referred to henceforth as the *best choice*.

In the situation described above, assuming that the cell that is particularly loaded is the *best choice*, it is clear that forcing a user to operate on a different cell means forcing that user to transmit with higher power. In a CDMA system, this causes quality to deteriorate for **all** users. The high power transmitted by the mobile terminal forced to operate on another cell, in fact, would create significant interference for the *best choice* station, and would thus raise the latter level of interference, which is already high because of its heavy load. As this example is intended to illustrate the substantial difference in radio resource management in a CDMA system and in an FDMA/TDMA system, we have intentionally neglected the macrodiversity function, which is important in managing radio resources in the case considered here. If the new user is located near the border of the cell in high load conditions[1], it could be possible to minimise transmitted power by setting up the call under macrodiversity conditions, or in other words by involving other radio base stations. If this is not possible, however (e.g. if the new user is not located near the cell border), it might be better to refuse to grant the new user access to the network so as not to jeopardise the quality perceived by all other users. This is the purpose of the *Admission Control* function described below.

The foregoing considerations assume an even greater importance in view of the fact that that UMTS will provide a plurality of services, each characterised by a certain bit rate and a certain level of service. Clearly, then, different services will call for markedly different transmitted powers, with obvious consequences: for example, poor management of a user who requires a 2 Mbit/s service could easily damage many users making use of a voice service.

system based on the CDMA technique. As the relevant resource in a CDMA system is the transmitted power, poor management of the latter can have a negative impact on the mobile radio system overall capacity.

In the following sections, these observations will be applied to the cell selection/re-selection, Admission Control and soft hand-over procedures.

(1) It should be borne in mind that it would be incorrect to speak of a congested cell in this context, as the CDMA system is characterised by *soft capacity*; in addition, the size of the cell varies according to load as illustrated in Chapter 3.

Cell selection/re-selection

In a cellular mobile system, each mobile terminal in the idle status (i.e. on but with no RRC connection set up) is associated with a radio base station. It is with this station that the mobile terminal will attempt to set up a call as soon as the user so requires. When it is turned on, the mobile terminal chooses a cell using the *cell selection* process. This cell will be updated through the *cell re-selection* process at frequencies which depend on changes in the mobile terminal location, the propagation conditions and/or other parameters. Consequently, it is clear that in a system based on the CDMA technique, it would be advisable to ensure that the mobile terminal will be associated with the *best choice* radio base station in order to minimise the power transmitted in setting up a call. This, however, may be difficult to accomplish. Therefore, it may be acceptable for the mobile terminal to exchange signalling messages with a radio base station other than the *best choice* station, providing that the network is capable of redirecting the call to the *best choice* station.

Admission control

The policies used for radio Admission Control (there is also such a thing as Network Admission Control, though it is beyond the scope of this chapter) limit access to the network even when radio resources are available. These policies are characteristics of access techniques which, like CDMA, feature soft capacity, i.e. where quality drops gradually as the number of calls increases. Maximum load is thus established by the operator, who consequently denies access to further calls in order to maintain quality at acceptable levels. These policies are not normally used in systems based on FDMA/TDMA techniques, which feature *hard capacity*: available resources (in terms of frequencies and time slots) are known beforehand. Until they are used up, it will always be possible to accept new users without substantial effects on perceived quality (providing that the 'frequency planning' process, which is not necessary for CDMA systems, has been carried out correctly).

Close-up 4.7

Admission control policies can be based on a variety of criteria. In this Close-up, we will focus on the interference criterion, whose importance has been emphasised on a number of occasions in these pages.

When a new user enters the network, power is transmitted on both the up-link and the down-link, increasing the interference perceived by other users. Consequently, accepting a new user is useless and counterproductive if the interference he produces is excessive.

On the up-link, the new user would cause an increase in the level of power received by the radio base station. If this increase, which is a function of the requested service, the current load and the channel conditions, is high enough to create problems in decoding all other users, it is best to deny the new mobile terminal access to the network.

On the down-link, where orthogonal spreading codes are used, the limit is usually set by the maximum power that can be transmitted by the radio base station (i.e. by a physical limit): in other words, the radio base station may not be in a condition to transmit the power needed by the new user which, as for the up-link, depends on the requested service, the system load and the channel conditions. On the basis of these considerations, an admission control policy could attempt to evaluate *a priori* (i.e. before the mobile terminal sets up the call) whether the power levels that would be necessary are compatible with the limits established by equipment and by the operator. If they are not, it could be advisable to deny the new user access to the network. Alternatively, the services requested by the new user could be renegotiated, as could the services provided to other users who are already making calls. For instance, reducing transmission rates to certain users would make it possible to reduce overall transmitted power, facilitating the entry of the new user.

Soft hand-over

As discussed in Chapter 2, soft hand-over is closely connected to macrodiversity, and the terminology usually adopted tends to confuse these two concepts. The advantages in terms of transmitted power savings, and hence in the system overall capacity, were illustrated in the Close-up at the end of Section 4.2.

The soft hand-over process consists of a set of functions: performing measurements, interpreting them, and transmitting the measured and interpreted data to the network entity in charge of controlling the process, as well as executing the soft hand-over algorithm *per se* and carrying out the actual hand-over operation.

4.4.4 Radio protocols and support for data and multimedia services

Data and multimedia services are essential to the growth of a third-generation system such as UMTS. Indeed, an important feature of

Close-up 4.8

As an example, one approach that can be used for the soft hand-over algorithm is as follows: if an appropriate reference signal received by a radio base station belonging to the Active Set (AS) is worse than the reference signal level received by the best station in the Active Set by a certain margin plus the associated hysteresis for a time period ΔT, the station in question is eliminated from the AS. If the difference between the reference signal level received by the best station in the AS and that received by a radio base station which does not belong to the AS is below a certain margin plus the associated hysteresis for a time period ΔT, the new radio base station is added to the AS if the latter is not full. If the AS is full but the reference signal received by a radio base station not belonging to the AS is better than that of the worst station in the AS by a certain margin for a time period ΔT, the worst station will be replaced.

UTRAN lies in the mobile terminal capability to manage different services simultaneously. Technically, this ability is supported by a series of characteristics at the radio protocol level.

- *Variable bit rate* on dedicated transport channels. This characteristic is particularly useful when the terminal must provide a set of different services. The potential for providing variable bit rates on a dedicated physical channel makes it possible to optimise resource usage also in these cases.

- *Multiplexing* different logical channels on the same dedicated transport channel.

- *Multiplexing* different dedicated transport channels on the same physical channel.

- Up-link common channel, which can be used as a particularly efficient support in providing data services.

- Down-link channel shared by several users, which is particularly suitable for Internet applications.

References

[1] 3G TS 25.301, 'Radio Interface Protocol Architecture'.
[2] 3G TS 25.401, 'UTRAN Overall Description'.
[3] 3G TR 25.922, 'Radio Resource Management Strategies'.

5

UMTS Network Infrastructure

Antonella Napolitano, Andrea Calvi and *Ermanno Berruto*

5.1 UMTS *network architecture*

The overall architecture of the UMTS system is shown in Figure 5.1. The entire system can be divided into two main segments: the access network, called the UMTS *Terrestrial Radio Access Network* or *UTRAN*, as described in the previous chapter, and the switching and routing infrastructure, or Core Network, presented in this chapter.

The UMTS architecture consists of two network domains: the Circuit Switched domain, which centres on the MSCs (*Mobile Switching Centres*), and the Packet Switched domain, which centres on the GSNs (*GPRS Support Nodes*). The two domains thus rely on two separate and parallel backbones. The first backbone, based on

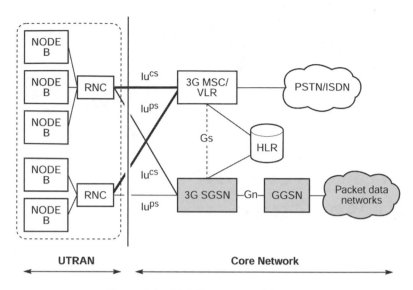

Figure 5.1 UMTS system architecture.

ISDN (*Integrated Service Digital Network*) derived technologies, carries voice traffic, while the second is based on technologies derived from the IP world and transports data traffic. The two domains are connected to the access network, which is shared by both types of traffic, through the I_u interface described in the previous chapter. As can be seen from Figure 5.1, this interface is in reality split into two parts: Iu^{CS}, which connects the access network to the circuit switched backbone, and Iu^{PS}, which connects the access network to the packet switched backbone.

It should be emphasised that the circuit switched UMTS backbone is derived directly from the classic GSM network infrastructure, whereas the packet switched UMTS backbone derives from the infra-structure used to introduce GPRS (General Packet Radio Service) in the GSM network. In fact, while the UMTS access network is entirely new and separate from that used for GSM, the core network infra-structure is a direct evolution of the GSM infrastructure.

In the early stages of UMTS service rollout, GSM operators will thus be able to share the network infrastructure between 2G and 3G access networks.

In view of the architectural choices and the direct derivation from GSM and GPRS, this chapter will deal separately with the two backbones making up the UMTS network infrastructure, and will present the main features of the GSM and GPRS networks on which these backbones are based.

For each domain, the architecture, the main signalling procedures and the most important innovations introduced in the process of evolving towards UMTS will be described.

Finally, the chapter will discuss the major trends for future versions of the system, which could make it possible to eliminate the current division of network domains and introduce an all-IP solution for voice and data traffic alike.

5.2 Circuit switched backbone

There can be no doubt that the circuit switched part of the UMTS architecture will inherit many of the features of today's GSM network platform.

Though it is true that the major technological innovations will be incorporated in the packet switched IP nodes, the circuit switched part will nevertheless have a fundamental role, as it will be responsible for providing voice services as in the present-day GSM network. It is not yet clear whether the UMTS circuit switched platform will also be used to support multimedia services or integrated voice-data services. In the light of the explosive success enjoyed by the Internet in recent years, however, we can reasonably expect that innovative services (i.e. services other than voice) will chiefly use the IP platform. A number of current trends appear to indicate that voice services could also migrate to IP in advanced versions of UMTS.

As most of the architectural structures for voice services used in second-generation mobile systems will be carried over to UMTS, a thorough knowledge of the basic principles and features of the GSM network is probably the best springboard for an analysis of the new systems.

5.2.1 Overview of the GSM network

Network architecture

The GSM network's basic switching architecture is shown in Figure 5.2 (white components). Main network elements are as follows:

- BSS (*Base Station Subsystem*), which is responsible for managing radio resources and for the interface between the radio channels and the physical link channels used to transmit information on the ground.

- MSC (*Mobile Switching Center*), This is the next-higher node in the network hierarchy, and is responsible for a group of physically adjacent BSSs. It acts as the system's nerve centre, controlling call signalling and co-ordinating the hand-over procedures between BSSs (or in the same BSS) which are triggered when a mobile terminal changes physical location.

The GSM architecture which controls terminal mobility is represented by the darker nodes in Figure 5.2 and consists of two network elements or databases:

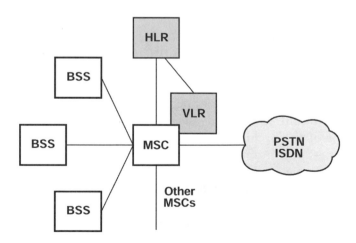

Figure 5.2   GSM network architecture.

- HLR (*Home Location Register*), which contains and updates information about the position of mobile terminals in terms of VLR area (or in other words, the identifier of the geographical area controlled by a specific VLR). In addition, the HLR stores information regarding user profiles and identification-authentication parameters.

- VLR (*Visitor Location Register*), which manages more detailed information about the terminal's position, tracking its movements at *Location Area* level (i.e. at the level of a subsection of the area managed by a VLR whose size can range from a few radio cells to the entire area covered by a BSS). In addition, authentication and identification parameters are stored locally by the VLR.

The connection between the switching platform and the mobility control architecture is situated at the level of the MSC nodes, which contains a module capable of requesting the location of a mobile terminal and the user profile. This information is needed in order to manage a mobile-originated or mobile-terminated call.

Manufacturers usually supply the MSC and VLR network nodes in a single integrated product, even though their functions are separate and specific; in practice, then, the two entities are implemented as a single physical element.

Signalling and switching issues

The overall architecture of the GSM network is similar to that of the ISDN network: calls are routed to and from the mobile network using the E.164 numbering plan, and the *Mobile ISDN number* (MSISDN) in any given country must comply with the ISDN numbering plan.

As in fixed ISDN network switches, moreover, a 64 kbit/s PCM (*Pulse Code Modulation*) circuit on each link between the MSCs and between the MSCs and the gateway node to other ISDN/PSTN networks is used for each call which is set up.

Unlike the ISDN network, a 16 kbit/s circuit is used at the interface between an MSC and a BSS for each call. This 16 kbit/s value results from GSM voice coding: to save scarce resources on the radio interface, voice is coded at 13 kbit/s rather than the classic 64 kbit/s of PCM coding. The lower coding rate is maintained by the mobile terminal up to the MSC, where data is re-coded in a 64 kbit/s stream compatible with the PSTN's PCM coding.

Like the telephone network, the GSM architecture's signalling network uses Common Channel (SS7) protocols and services. This protocol stack was used as the basis for developing the MAP (*Mobile Application Part*) protocol which is specific to GSM and handles dialog between HLR, VLR and MSC for managing mobility and authentication procedures.

Main mobility management procedures

This paragraph describes the main mobility management procedures used in the GSM network.

Location updating procedure

In order to be reached by an incoming call, the mobile terminal connected to the GSM network must inform network intelligence (or in other words, the functions managing the information contained in the HLR and in the VLR) of its location.

This is accomplished by means of the *Location Updating* procedures, in which information about the Location Area and the VLR area is sent to the VLR and HLR, respectively.

Initiated by the mobile terminal, the procedure can be triggered by a change in the terminal's location or activated periodically.

Paging and routing procedures

In the case of an incoming call for a mobile subscriber, the GSM network is requested to reach and connect the mobile terminal using its MSISDN. Each MSC in the network is potentially capable of accepting this request from an external network. As the HLR is the only network element which knows and manages the location of all active mobile terminals, the MSC asks this node for the VLR area in which the terminal can be found. If this initial part of the procedure (which takes place if the mobile terminal is on and active) is concluded successfully and an MSC/VLR address is determined, the call is routed to the MSC/VLR currently associated with the user.

At this point, the VLR is requested to retrieve detailed information regarding the mobile terminal's position, i.e. its Location Area. A paging message is broadcast in the Location Area thus identified which contains the called user's identifier. This identifier permits the other mobile terminals located in the same Location Area to ignore the request, and enables the called user's terminal to understand that the call is directed to it.

If the mobile terminal is in the coverage area and the user is allowed to take the call (e.g., the network has sufficient resources available) a radio channel is dedicated to the terminal and the call is routed via the appropriate BSS to the radio cell serving the user.

5.2.2 UMTS CS network architecture

As can be seen from Figure 5.3, the architecture of the UMTS network is similar to that of the GSM. Following the introduction of the new radio interface, the BSS was replaced by UTRAN, which was described in Chapter 4. The UMSC (*UMTS MSC*) is now provided

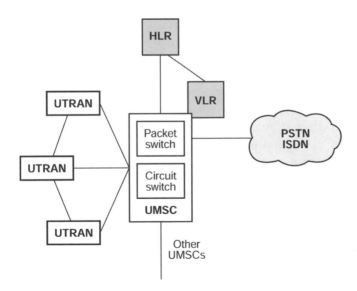

Figure 5.3 Circuit switched UMTS network.

with both circuit and packet switching capabilities, though the main interfaces and control architecture remain the same.

5.2.3 Innovative features with respect to GSM

The major difference between second-generation mobile systems such as GSM and the third-generation UMTS system lies without doubt in the adoption of the new radio interface. First of all, this interface will make it possible to reach higher bit rates, ensuing greater flexibility. Though it is true that the highest bit rate that can be attained with a W-CDMA interface is 2 Mbit/s, a wide range of intermediate rates will also be available, so that in theory it will be possible to support any type of present or future service. This is a decisive change with respect to the GSM system, where the only channel additional to the voice channel is the 9.6 kbit/s data channel, which is often unable to ensure fast and efficient information transmission. Furthermore, the UMTS circuit switching infrastructure must be capable of setting up, maintaining and releasing circuits at different speeds, with a level of complexity significantly above today's rigid 64 kbit/s switching architecture.

From the beginning, the UMTS transport architecture must be able to provide its services to both the circuit switched part of the platform and to the IP-based part. By contrast, GPRS in second-generation systems was conceived as an addition to GSM; as a result, its transport network was separate from that used to transport voice. In UMTS, telephony and data transmission were integral parts of network architecture objectives from the outset, and it was immediately clear that it was necessary to adopt a transport technology which was optimised for both.

The need for a new transport and switching architecture was thus recognised during the UMTS specification process.

Use of ATM

When a transport technology is needed which emulates voice transmission circuits and at the same time permits data transfer, the choice necessarily falls to ATM.

ATM, in fact, is already one of the most widely used transport technologies for IP wide area networks (and thus performs the role of a Data Link layer protocol). Thanks to an enormous range of standardised performance features and protocols for call routing and addressing, moreover, it can incorporate the signalling, transport and networking capabilities typical of telephone networks, a fact which sets it apart from other IP transport protocols such as Ethernet, Fast Ethernet, Frame Relay and FDDI. The reason ATM can satisfy such disparate needs with such a high level of flexibility is that it was designed from the outset to work in a multiprotocol environment: to enable ATM to support highly dissimilar services, *ATM Adaptation Layers* (AALs) differing widely in their characteristics were specified.

Two AALs were included in the transport platform UMTS services: AAL5 and AAL2.

AAL5 is the adaptation layer which makes it possible to transport an IP packet (whose length can by definition vary up to a maximum of 65536 byte) in a series of ATM cells of fixed length (53 byte, including 5 header bytes and 48 payload bytes).

The main functions of AAL5 are thus segmentation and re-assembly: an IP packet is divided into smaller AAL5 PDUs, which are the payload of the ATM cells (this operation includes inserting a number of filler bits if the size of the IP packet is not an even multiple of 48 bytes), and the divided packets are transmitted and reassembled at the receiving end.

AAL2 performs a fundamental role in UMTS, and in the access network in particular. A detailed description is provided in Chapter 4.

From GSM hand-over to UMTS streamlining

Hand-over, as we know, is the mechanism which makes it possible to cope with the user's movement during a call by dynamically changing the radio base station with which the mobile terminal is connected. In the network architecture's evolution from GSM to UMTS, this procedure has been extensively revised thanks to the introduction of the I_{ur} interface.

In GSM, in fact, hand-over between different BSCs involved the MSC through an operation which required switching the voice path from the old MSC-BSC interface to the new one (Figure 5.4). This

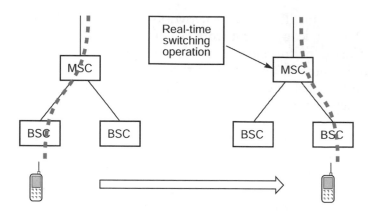

Figure 5.4 Hand-over between BSCs in the GSM system.

operation had to be performed in real time in order to reduce the impact of hand-over on user-perceived quality.

In UMTS, thanks to the use of the I_{ur} interface, the access network (UTRAN) can manage hand-over independently without involving the MSC. The procedure is thus faster and simpler, as it is not necessary to synchronise the MSC switching matrix with radio base station movement (Figure 5.5). The initial effect of UMTS hand-over is thus to 'extend' the UMSC-RNC interface via the I_{ur} which makes it possible to reach the new RNC. Only subsequently, with no need for real time operation, will the connection to the new RNC be set up and the I_{ur} extension released.

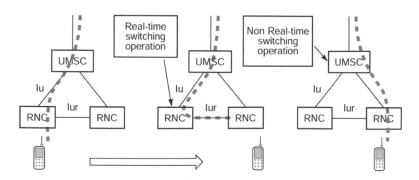

Figure 5.5 UMTS streamlining for RNC changes.

The I_u interface redirection procedure, also known as streamlining, is designed to optimise resources (or in other words to minimise the number and length of terrestrial circuits used and thus avoid forming a 'chain' of RNCs) once hand-over has been consolidated.

The streamlining procedure coincides with the SRNS (*Serving Radio Network Subsystem*) relocation procedure (see Chapter 4), which moves terminal control intelligence from the old RNC to the new one.

This operation's delay relative to terminal movement prevents the MSC from being affected by 'oscillatory' phenomena, or in other words small terminal movements (or changes in radio interface conditions) which can trigger hand-over in one direction followed by immediate return, also known as the ping pong effect.

Transcoder location and switching in the transport platform

In all mobile radio systems, radio bandwidth is the scarcest, and thus the most precious, resource. For this reason, voice coding on the radio interface must ensure maximum savings, or in other words, the highest possible compression factor.

While GSM voice coding produces a 13 kbit/s data stream, the new type of codec used for UMTS makes it possible to contain variable bit rate in the 4–13 kbit/s range.

To permit mobile radio interconnection with the world of fixed networks (where transport is based on 64 kbit/s PCM voice coding), low bit rate coding must be transformed into PCM coding (with a consequent increase in bit rate) as described in the foregoing paragraphs.

The maximum possible savings in terms of link occupation would be achieved by maintaining 4–13 kbit/s voice coding in the entire mobile radio network infrastructure, or in other words by positioning the transcoders at the edges of the network (Figure 5.6).

This configuration would permit major transport cost savings by comparison with the GSM networks, where transcoding takes place at the MSC and the entire transport network between switching centres uses 64 kbit/s circuits.

Clearly, there is a price to be paid for this optimisation: low bit rate transport can take place only on ATM with AAL2, and the entire

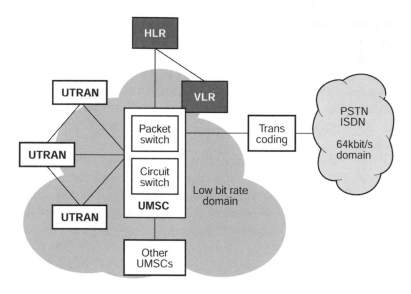

Figure 5.6 Transcoder location.

PCM-based transport and 64 kbit/s matrix-based switching structure would have to be changed.

UMTS call control for multimedia services

As the data transfer speeds permitted by the new radio interface could in theory range from a few byte/s up to 2 Mbit/s, conventional services such as telephony and new multimedia services such as video, web browsing and message or image exchange will be available via both packet switching and circuit switching.

For this reason, UMTS Call Control (or in other words the capability which supervises such aspects as number analysis, called-party ringing and disconnection at the end of the call) requires features additional to those provided in GSM, where the only permitted component was voice.

In the new system, it will be necessary to manage two or more components (or media) during a call. For example, it will have to be possible to initiate a phone call and then add video.

There were two possible ways to achieve this end: developing a new standard for multimedia and mobile telephony, or reusing existing standards.

For the sake of simplicity, it was decided to adopt the second approach and, in particular, a combination of GSM Call Control and H.324, the family of standards which are commonly used on the Internet to manage multimedia applications. Thus, GSM Call Control will make it possible to 'open a link' between caller and called party as takes place in a phone call today, while H.323 will make it possible to add and remove the various media.

5.3 Packet switched backbone

The packet switched part of the UMTS architecture will inherit many of the features of the current GPRS network platform.

Developed by ETSI, the GPRS service is the standard which permits packet data transmission to be introduced in the GSM system. GPRS has also been accepted by the United States' TIA (*Telecommunications Industry Association*) as the data standard for the TDMA/136 system.

GPRS permits packet mode data transmission and reception on both the radio interface and the network infrastructure without employing circuit switched resources.

By adding GPRS to the GSM network, the operator can thus offer efficient radio access to external IP-based networks such as the Internet and corporate Intranets.

5.3.1 Overview of the GPRS network

Network architecture

The classic GSM network does not provide sufficient capabilities for routing packet data. For this reason, the conventional GSM structure has been extended (Figure 5.7) by introducing a new class of logical network entity called GSN (*GPRS Support Node*).

The GSN nodes manage interconnection with the other networks and perform a variety of functions, including subscriber management,

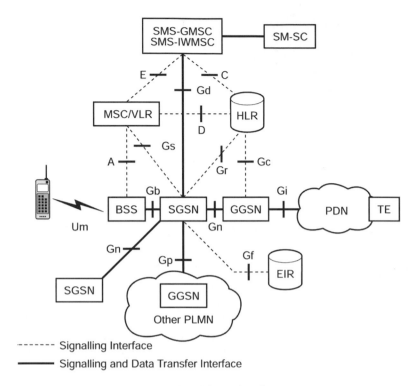

Figure 5.7   GPRS logical architecture.

billing and security, mobility management, roaming and geographic
re-routing, virtual connection control, and packet transmission. The
Serving GPRS Support Node (SGSN), which is connected to the access
network and is at the same hierarchical level as the switching centres
(MSCs/VLRs), is the node that serves the GPRS mobile terminal,
retaining location information and carrying out functions related to
communication security and access control. The *Gateway GPRS
Support Node* (GGSN) is seen from outside as the access port to the
GPRS network and acts as an inter-working unit *vis-à-vis* the external
packet switched networks. Within the network, the GGSN is
connected to the SGSN nodes by means of an IP-based transport
network. The HLR database must be updated with new functions
for storing data about GPRS users' subscription profiles and routing
information. Finally, the *Short Message* service centres (SM-SCs) are

enhanced so that SM transmission is also possible via the GPRS radio channels.

The coverage area, i.e. the portion of the territory in which the service is guaranteed, is logically organised in location zones which enable the network to know where the mobile terminal is located during its movements. These zones, which are similar to the Location Areas (LAs) used by the circuit switched network, are called *Routing Areas* (RAs) and are smaller than the LAs.

To satisfy the needs of different market segments, the GPRS standard specifies three different kinds of mobile terminal:

- *Class A terminal*: the terminal can be connected simultaneously to the classic MSC network in order to use circuit switched services, and to the GPRS network in order to transmit and receive packet data. In other words, this class of terminal can use packet and circuit traffic simultaneously.

- *Class B terminal*: the terminal can be registered simultaneously on both the circuit and packet switched networks, but cannot send and receive traffic in the two modes simultaneously.

- *Class C terminal*: the terminal can be registered for either packet mode or circuit mode, and can thus support only traffic for the type of service for which it is registered.

Routing and signalling issues

The network infrastructure for implementing the GPRS service is based on IP technology. Using this technology for transmission to and from mobile terminals requires special routing solutions. In fact, the IP version used in the GPRS standard does not contemplate a mobility management mechanism. Consequently, a specific routing method which will be briefly illustrated below was introduced in the GPRS standard.

For data packet transmission in the GPRS network, the mobile terminal is identified by an IP address assigned to it either permanently or dynamically at the time the session is set up.

Packets arriving from the external networks are delivered to the GGSN in the GPRS network to which the mobile terminal belongs. The GGSN has the routing information needed to send the packet (using the tunnelling method described in Close-up 5.1) to the SGSN network entity serving the geographical area where the mobile terminal is currently located. In turn, the SGSN sets up a logical link with the mobile terminal through which the packet is delivered.

In the case of a mobile-originated transmission, the SGSN encapsulates the incoming packets and transfers them to the reference GGSN, where they are forwarded to the destination data network.

All data about GPRS subscribers which the SGSN node needs in order to route and transfer data are stored in the GPRS register, which conceptually is part of the GSM system's HLR node. The GPRS register contains routing identification and the match-up between the subscriber identifier (IMSI – *International Mobile Subscriber Identity*) and the assigned IP address, and between the latter and the reference GGSN.

GPRS has a series of parameters which characterise data transmission in each active context. The main parameter affecting how IP packets are transferred is quality of service. This parameter is defined through the following attributes:

1   Precedence: indicates how the service is prioritised among various users. Three levels are available: high, medium and low.

2   Delay: refers to the end-to-end delay in transmitting a packet from origin to destination. Three classes with an established maximum delay are specified, plus a best effort class.

3   Reliability: a service's reliability class differs according to its correction capacity and fault tolerance.

4   Throughput: indicates the throughput required for the user. Two characteristics of this parameter can be negotiated: maximum bit rate and average bit rate.

Close-up 5.1 – Tunnelling mechanism in the GPRS network

The tunnelling function enables data flows, to which addressing and control information have previously been assigned, to be tunnelled through two points of a network or through different networks. In GPRS, this mechanism is used to manage user mobility and thus the network movements of the IP addresses assigned to the users. The normal routing functions in an IP network are static in nature, as packets containing a given address are always sent to the same destination. For GPRS, on the other hand, packets intended for a certain user must be routed to that user's current position, which can usually change. In order to solve this problem, the GGSN 'encapsulates' each IP packet, entering the GPRS network, in another IP packet containing the address of the SGSN node controlling the mobile terminal at the time the packet enters the network. In this way, all packets addressed to the users controlled by a certain SGSN are transferred to that node, creating 'tunnels' between GGSN and SGSN. When the user changes location in the network and comes under the control of another SGSN, the GGSN changes the encapsulation address, thus varying the destination of the tunnel opened for that particular user. During all of the user's movements in the network, there will be an active tunnel for the current data session, which opens from the GGSN and exits in the SGSN controlling it at the time.

Main control procedures

The main control procedures used in the GPRS network are described in this paragraph.

GPRS attach and context activation procedures
Before a mobile terminal can access GPRS services, it must inform the network of its presence by performing a *GPRS Attach* procedure (Close-up 5.2) to the SGSN node. The Attach procedure entails: updating location information in the HLR; transferring information

from the old SGSN, where the mobile terminal was formerly registered, and the new SGSN; and deleting data from the old SGSN (as
well as from the old VLR if the mobile terminal was also registered
with the GSM network for circuit switched services).

To transmit or receive data, an MS must then activate a PDP
context (see Close-up 5.3). Activation of a PDP context informs the
reference GGSN that the mobile terminal is present and makes it
possible to transfer data packets to and from the corresponding

---

Close-up 5.2 – GPRS attach procedure

The GPRS Attach and PDP (Packet Data Protocol) Context Activation procedures must be performed in order to permit the GPRS
user to connect to the external data networks. To all intents and
purposes, the GPRS Attach procedure, like the corresponding
procedure in the circuit switched world (*IMSI Attach*), is used to
ensure that the network is informed of the mobile terminal's
presence. Once the terminal is registered, the network knows its
location at Routing Area level and its service characteristics.The
GPRS Attach procedure consists of performing the following steps:

1   The mobile terminal requests the network to activate the proce-
    dure. The request that the terminal sends to the SGSN indicates
    the terminal's capability to handle high transmission speeds
    (simultaneous use of several radio interface time slots), the
    encryption algorithm used and the type of mode (circuit, packet
    or both) for which registration is requested.

2   The authentication procedure is performed.

3   Subscription data are transferred from the HLR register to the
    SGSN and MSC/VLR nodes.

4   The SGSN node informs the mobile terminal that the requested
    procedure has been successfully completed.

user. The PDP context contains specific addressing information for packet transfer. For each PDP context, the mobile terminal can be assigned a static address (ETSI X.121, IETF IPv4 or IPv6 type) established at the time of subscription, or a dynamic address allocated at the time the GGSN activates the PDP context by the operator of the user's home network (HPLMN – *Home Public Land Mobile Network*) or of the visited network (VPLMN – *Visited Public Land Mobile Network*). Dynamic addresses only allow data transfers originated by the mobile terminal.

For mobile-terminated calls, if the GGSN receives packets before the mobile terminal has activated a PDP context, it can start a network-originated PDP context activation procedure (such a procedure is possible only in the case of statically assigned addresses).

When the detach procedure is requested (by the network or by the mobile terminal), all PDP contexts for a given terminal are deactivated. The detach procedure can also be originated implicitly when a predetermined time expires during the period in which there is no mobile terminal activity (e.g., no data sent or received).

For roaming subscribers who have a PDP address allocated by the HPLMN, a forwarding path between the HPLMN and the VPLMN is created for communicating with the mobile terminal in both directions. Protocols such as BGP (*Border Gateway Protocol*) (IETF RFC 1771) can be used between the BG (*Border Gateway*) routers on the basis of bilateral agreements between the operators.

As indicated earlier, IP addresses can be assigned to mobile terminals statically (a given user is always assigned the same IP address, which is stored in the HLR register) or dynamically (the assigned IP address changes at each context activation).

The term transparent access is used in cases where the mobile terminal is assigned an address from the mobile radio network operator's addressing space. In such cases, no security mechanism is envisaged when the PDP context is activated, given that authentication and encryption can only be performed on an end-to-end basis (e.g. by means of a protocol such as IPsec).

The access is defined as 'non-transparent' when the mobile terminal is assigned an IP address from the addressing space of a connected ISP (*Internet Service Provider*) or Intranet. In this case, the mobile terminal must perform the authentication procedure

Close-up 5.3 – PDP context activation procedure

In order for the mobile terminal to communicate with external data networks, the context for the packet data transfer protocol must be activated through a procedure called *PDP Context Activation*. The PDP Context describes the characteristics of the link with the external data network, viz.: type of network, destination address, the address of the GGSN to be used and the quality of service characteristics.

1   The mobile terminal requests PDP context activation, specifying a number of parameters, including whether address assignment is to be static or dynamic and the required quality of service.

2   The SGSN node validates the request on the basis of the subscription data received by the HLR register at the time of registration.

3   The SGSN determines the GGSN node's address on the basis of the information provided by the mobile terminal and the subscription data.

4   A logical link – called a GTP tunnel – is created between SGSN and GGSN.

5   The SGSN node asks the GGSN node to allocate an IP address and transfers it to the mobile terminal.

6   At this point, communication between the mobile terminal and the external data network can begin.

towards the GGSN at the time the context is activated, while the GGSN must in turn request authentication on behalf of the user from an ISP or Intranet server. Examples of protocols which may be used to guarantee user authentication on the GGSN-ISP/Intranet link include Radius and DHCP (*Dynamic Host Configuration Protocol*). At the end of this operation, the mobile terminal is seen as a user of the network to which the assigned address belongs, and the packets containing this address will thus travel through this network before arriving at the GGSN.

GPRS mobility management

The mobile terminal knows its location in terms of visited cell and RA (Routing Area). In the network, the terminal's location is tracked on two levels, depending on the status of the mobility management procedure. When the mobile terminal has carried out the Attach procedure for the GPRS network, but is not involved in an active connection, the network tracks its movements at RA level (Close-up 5.4). When the mobile terminal is involved in an active connection, its location is tracked at cell level.

Mobility between SGSN and GGSN is managed by means of the *GPRS Tunnelling Protocol* (GTP). The GTP protocol also permits information transfer between two or more SSGNs at the time the mobile terminal changes SGSN.

To keep its location with the network up to date, the mobile terminal performs a mobility management procedure when it enters a new cell or a new RA. Updating the RA may entail changing SGSN (inter-SGSN mobility); in this case, a procedure is activated which involves:

- The old SGSN, in order to transfer PDP context information about the active contexts and to set up a forwarding path for the data which are still in transit between the GGSN and the old SGSN.

- The GGSN at each active PDP context, in order to update the GTP tunnels.

- The HLR, in order to store the new SGSN information and remove the information concerning the old SGSN.

A critical element in transporting real time services such as voice over IP on GPRS is RA updating. This procedure must be fast enough to permit terminal mobility without breaks in the service. In GPRS, in fact, the concept of hand-over is implemented in terms of cell reselection and RA updating.

Close-up 5.4 – Routing area update procedure

The *Routing Area Update* (RAU) procedure is carried out when the GPRS mobile terminal relocates to another routing area. This area, defined independently of the GSM Location Area organisation, makes it possible to distribute messages on the basis of specific service criteria.

The mobile terminal recognises a change in Routing Area by reading the contents of the broadcast signalling channel transmitted by the radio base station which guarantees radio coverage for the cell involved. This channel, in fact, transmits the identifier for the Routing Area to which the cell belongs. When cell selection changes, the mobile terminal checks the transmitted identifier and, if it differs from that of the previous cell, the terminal starts the Routing Area Update procedure which updates the location information for the user in the network registers.

Co-ordination between packet and circuit modes

Co-ordination between packet and circuit modes is accomplished by means of an interface called G_s (Figure 5.7) whereby an association between MSC/VLR and SGSN can be created. This association is particularly useful in cases where it is necessary to manage mobile terminals that can connect simultaneously to circuit switched services and packet switched services (Class B terminals).

The association is created between the two entities when the SGSN node serving a particular mobile terminal knows the address of the MSC/VLR on which the same terminal is simultaneously registered and vice versa. One of the advantages of the G_s interface is that it limits the signalling traffic on the radio interface. When a mobile terminal is connected to an SGSN, in fact, it can if necessary carry out a combined Location Area Update and Routing Area Update procedure to update its location for both circuit and packet switching modes. In addition, the paging message for an incoming circuit switched call can be forwarded via the G_s interface to the SGSN node and thence to the mobile terminal.

The *Combined LA/RA Updating* procedure is as follows: when the mobile terminal moves into a new RA, it sends an RA *Update Request* message to the SGSN. If the LA has also been changed at the same time, the SGSN node will send an *LA Update Request* to the VLR. Location information will thus also be updated for the circuit switched domain without having to involve the mobile terminal directly.

As the RA is usually smaller than the LA and at most will be the same size, it is assumed that the number of RA updates will be greater than the number of LA updates, while a change in LA will invariably entail a concurrent change in RA.

In the case of the paging procedure, if the MSC/VLR node needs to forward a paging message for circuit switched services to a mobile terminal registered on the GPRS network, it will send this message to the SGSN via the G_s interface. The SGSN will then transmit the paging message in the RA where the mobile terminal is located.

5.3.2 UMTS packet switched network architecture

The structure of the packet switched part of the UMTS network is similar to that of the GPRS, where the BSS access segment is replaced by the UTRAN access network based on W-CDMA (Figure 5.1). Connection between the Core Network and UTRAN is guaranteed by a new interface called I_u, which specialises in managing both the packet switched component and the circuit switched component (see Chapter 4).

5.3.3 Innovative features with respect to GPRS

The standards-writing organisations are now taking steps to adapt the GPRS backbone to third-generation systems, starting from the GPRS architecture and the mobility management protocols described above. Two areas are particularly critical: mobility management and quality of service control. In both of these areas, contributions from the world of information technology are becoming increasingly important. Using an IP backbone such as the GPRS version, in fact, makes it possible to introduce mechanisms specified by IETF (*Internet Engineering Task Force*) work groups in the UMTS network's packet switching segment. Naturally, applying these mechanisms to a UMTS context will require establishing a good level of inter-working with the similar procedures, typical of radio telecommunications, which are used in the UTRAN access network. The new mechanisms that are now being considered for the UMTS network will be briefly described below.

IP mobility management (Mobile IP)

Studies are currently addressing the use of the *Mobile IP* protocol (MIP) (Close-up 5.5) to manage mobility in the UMTS network. As indicated in Chapter 2, Mobile IP is a mechanism introduced to guarantee computer mobility between different IP networks. As the packet switched UMTS backbone is based on IP technology, an attempt is being made to provide inter-working between the classic GSM mobility procedures used in the GPRS, and those specified for Mobile IP.

Three successive stages have been outlined for introducing the Mobile IP mechanism in the UMTS network.

- Stage 1: this is the minimum configuration for an operator who wishes to provide the Mobile IP service. The current mechanisms are maintained for managing mobility within the UMTS network, while Mobile IP is used to manage roaming between different systems (e.g. local area networks) and UMTS, without losing the work session in progress.

- Stage 2: the Mobile IP mechanism could be used in the UMTS network for mobility between different GGSNs. To achieve more efficient addressing, the mobile terminal could – as a result of a hand-over – change the GGSN possessing *Foreign Agent* (FA) capabilities. In this case, the active PDP context and the associated forwarding address are updated using the Mobile IP mechanism only if data transfer is not in progress. This mechanism is particularly useful if there are a number of GGSN/FAs or when the GGSN and SGSN nodes are co-located. In the case of data transfer, the mobile terminal moving between the old and the new SGSNs could keep the PDP context active for the old GGSN and, as soon as transfer is complete, associate the new SGSN with the new GGSN/FA.

- Stage 3: SGSN and GGSN could be integrated in a single node, thus collapsing the associated G_n interface, but retaining all of the other interfaces shown in Figure 5.7 without change. In this case, the Mobile IP mechanism could be used both within the UMTS network for mobility between different SGSN/GGSNs, and between different networks. In such situations, the Mobile IP mechanism could also manage hand-over with data transfer in progress.

Close-up 5.5 – Mobile IP

Mobile IP is a protocol which allows mobile computers to move freely (i.e. to roam) in other networks while retaining the same IP address. Mobile IP consists of three elements: *mobile node, home agent* and *foreign agent*. While the last two elements are essentially routers with a few special properties, a mobile node is a mobile computer. The presence of a home agent in a network enables mobile computers to move in other networks. A foreign agent permits mobile computers arriving from other networks to visit the network in which it is located. A host which exchanges messages with a mobile computer is called a *corresponding node*. It may be an ordinary Internet host, or another mobile computer.

Mobile IP enables mobile computers to use two IP addresses effectively: one for identification (the *home address*), and one to forward traffic (the *care-of-address*). There are two possibilities for the care-of-address: it can be an address assigned temporarily to the mobile computer, or it can simply be the address of the foreign agent with whom the mobile computer is registered.

The mobile computer uses an *agent discovery* protocol to identify the foreign agent which is willing to provide mobility support in the network the computer is visiting. Foreign agents and home agents periodically transmit multicast or broadcast *agent advertisement* messages to indicate that they are present in the network; in addition, a mobile computer can ask that a message of this kind be sent by means of an explicit request called an *agent solicitation*. The mobile computer can discover the foreign agent's identity and care-of-address upon receiving an advertisement message. Once a foreign agent has been discovered, the mobile node notifies its home agent of the care-of-address and the registration validity period. The home agent completes registration by updating its addressing table and creating a mobility link which associates the mobile computer's home address with the care-of-address assigned to it temporarily.

When the IP packets sent by a corresponding node arrive at the network to which the mobile computer belongs, the home agent forwards them to the care-of-address by means of an encapsulation method which is also called tunnelling (see Close-up 5.1), using the care-of-address as the address of a new IP packet in which the original packet is inserted. The foreign agent, at which the tunnel ends, retrieves the original packet and sends it to the mobile computer.

In the opposite direction, the mobile computer simply sends its packets via a router in the visited network, given that transmission is independent of the origin address (Figure 5.8).

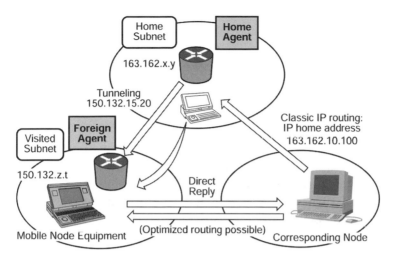

Figure 5.8 Mobile IP mechanism.

Quality of service in the UMTS network

End-to-end network services (i.e. between two terminals) are characterised by a certain Quality of Service (QoS) which is provided to the
user, who will thus have a personal perception of this quality. To
provide a given network QoS, it is necessary to establish a *bearer
service*, with specified characteristics and capabilities, from the service's source to its destination.

In defining QoS classes for the UMTS system, the restrictions and
limitations deriving from the presence of the radio interface must be
borne in mind.

Given the various error characteristics which are typical of the
radio interface, it would thus not be reasonable to define complex
mechanisms like those for the fixed network. The QoS mechanisms to
be provided in cellular networks must be robust and, at the same time,
capable of ensuring reasonable resolution.

The following QoS classes are envisaged for the UMTS system:

1   Conversational class,

2   Streaming class,

3   Interactive class,

4   Background class.

The main feature which makes it possible to distinguish between the various classes is their sensitivity to delay. Sensitivity is highest in the Conversational class, which is suitable for highly-sensitive traffic, and decreases class by class to reach a minimum in the Background class, which is practically insensitive to delay.

Accordingly, the first two classes are suitable for transporting real time traffic, while the Interactive and Background classes are designed for conventional Internet applications such as web browsing, e-mail, Telnet, FTP and News. Thanks to their less stringent delay constraints, the latter two classes provide better fault tolerance, using better channel coding and retransmission mechanisms than the Conversational and Streaming classes. Interactive class traffic has higher priority than Background traffic, which can use transmission resources only when they are not required for applications belonging to other classes.

Conversational class
This class is employed for real time conversations between users, like the conventional voice, voice over IP and videoconferencing services. In these services, transfer time must be kept low and, at the same time, the temporal relationship between the various data stream components must be maintained constant. In particular, the characteristics of these parameters are determined by human perception.

Streaming class
This class is employed in cases where the user wishes to watch (or listen to) real time video (or audio) streams. The transmission service is always uni-directional, from a network server to the user. As for

the Conversational class, the temporal relationship between the various data stream components must be maintained constant, though there are no special requirements for low transfer delay. The stream, in fact, is realigned by the receiving application, and the limits of this realignment mechanisms are much higher than the limits of human perception.

Interactive class

This class is employed in cases where the user requests data and interacts with a remote device. Typical applications include web browsing, database queries, access to network servers and measurement data acquisition. The main requirements for this class concern round-trip delay, given that the application requesting data will then wait for them for a predetermined time, and data integrity, i.e. a guaranteed low bit error rate.

Background class

Services in this class are those in which the user requests data files and waits for them to be received in a background process, which is thus secondary by comparison with higher-priority processes. Applications of this kind include e-mail and SMS transmission, database transfer and measurement data reception in background mode. For this class, the receiving application does not establish time limits for receiving the requested data, and is thus practically insensitive to delay. Data integrity, on the other hand, is extremely important.

Multimedia service management

In the UMTS network, new multimedia services have been designed for both the circuit switched part, which was presented in paragraph 5.2, and for the packet switched part.

In this connection, call control has been extended to include additional tasks that make it possible to manage the different media used during communication, e.g. passage from voice-only to video and back. The solution adopted for the packet switched part of UMTS is the H.323 multimedia call model. In particular, the H.323 standard is combined on the one hand with signalling developed for GSM voice control (see Section 5.2.3.) and on the other hand with GPRS session management (PDP Context Establishment). This decision reflects the desire to provide a multivendor environment between the mobile terminal and the network, as well as the desire not to rule out the possibility of selecting other protocols for future UMTS system developments.

It should be noted that the multimedia service management protocol is provided transparently, through the PDP context session established using typical GPRS signalling.

Figure 5.9 illustrates the relationships between multimedia protocols for both the circuit switched and packet switched parts. The *Multimedia Protocol* block indicates that the capabilities needed for multimedia service control in the packet switched part must be present in both the mobile terminal and in the network's H.323 Gateway. It is important to recognise that the Multimedia Protocol function manages both the control plane, i.e. the signalling stream, and the user plane, i.e. the data stream.

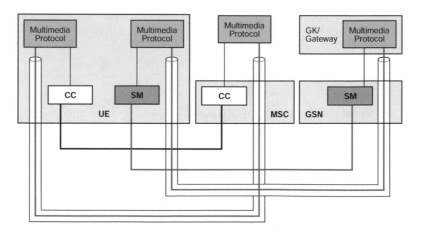

Figure 5.9 Multimedia service management.

5.4 *Future developments*

UMTS system specifications are continuously evolving through the definition and subsequent introduction of annual Releases. The description provided in the foregoing paragraphs applies to the Release 1999 (R99) UMTS network infrastructure completed and approved by 3GPP in 1999 and transferred to the respective regional standardisation bodies (ARIB/TTC for Japan, CWTS – *China Wireless Telecommunication Standard Group* for China, ETSI for Europe, T1 for the United States and TTA for Korea). At 3GPP , the possibility of specifying a UMTS network solution for Release 2000 (R2000) based entirely on IP technology is taking increasingly concrete shape. In this solution, the double backbone structure is abandoned in order to pass to a single network infrastructure in which voice traffic is also managed through packet switching. Work on specifying the UMTS R2000 network centres on the following assumptions:

1   The network architecture is based on IP packet technology in order to provide real time and non real time services simultaneously.

2   The network infrastructure is based on an evolution of the current GPRS network.

3   Mobile terminals are based on IP and service integration is achieved through IP.

4   The network must guarantee personal mobility and inter-working between the mobile and fixed networks for both the voice service and the data service.

5   The system must provide a voice quality comparable with or better than that ensured by current telecommunications networks.

6   The system must guarantee a level of network reliability comparable with that which is now provided.

7 The all-IP network must provide a minimum set of second-generation network services.

8 All IP-based interfaces and the interfaces associated with the network must be improved to guarantee support for real time multimedia services.

9 The system must provide separation between the service control procedures and the call and connection control procedures.

10 In the signalling network, today's common channel (SS7) transport is replaced by IP-based transport

11 The network architecture must be independent of OSI transport layers one (L1) and two (L2).

5.4.1 Network architecture

Initial studies have made it possible to delineate a reference architecture for Release 2000 as shown in Figure 5.10.

This architecture, which is profoundly innovative and is designed to provide integrated IP services, is achieved by improving the GPRS-based packet switched section of the UMTS R99 network. Call control is still supplied by the operator's network, but is now based on IP technology and is derived from SIP or H.323 type protocols.

Two possible access networks are envisaged: the UTRAN network defined by 3GPP (see Chapter 4), and one, called ERAN (EDGE Radio Access Network), based on EDGE (Enhanced Data rate for GSM Evolution) access, which introduces a new modulation scheme to increase the provided bit rate.

In the UMTS R2000 architecture, new network elements are added (Figure 5.10) in order to implement a single network backbone based on IP. These new elements are briefly described below:

• *Call State Control Function* (CSCF): carries out call control, service

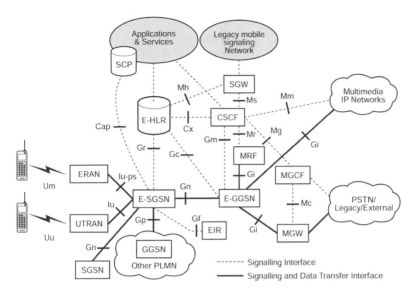

Figure 5.10   All-IP architecture.

switching, address translation and coding type negotiation functions.

- *Media Gateway Control Function* (MGCF): this is the interface
  point for signalling between the IP-based UMTS R2000 network
  and the circuit switched networks.

- *Media Gateway Function* (MGW): this is the interface entity for
  voice traffic between the IP-based UMTS 2000 network and the
  circuit switched networks.

- *Multimedia Resource Function* (MRF): carries out all functions
  needed in order to set up multiparty and conference calls.

- *Signalling Gateway Function* (SGW): performs conversion between
  common channel-based signalling (SS7) and IP-based signalling.

  Naturally, it will also be necessary to increase the capabilities of current UMTS R99 network elements such as the GSN and HLR nodes so that integrated services can be managed on IP.

The IP-based UMTS R2000 network will obviously have to guarantee full backward compatibility with the previous Release (R99).

5.4.2 Quality of service in packet switched networks

As noted above, one of the crucial aspects of a UMTS network solution based entirely on IP is that of quality of service (QoS) in terms of available bandwidth, total end-to-end delay, and jitter.

Traditionally, IP networks provide a single service class called *best effort*, which means that packets may be lost and delays are not controlled. Each IP packet, or *datagram*, is sent independently of all other packets: as available bandwidth and delays depend on the instantaneous status of the network, there are thus no guarantees on whether the datagrams will be delivered to the receiving station and on the resulting end-to-end delay.

While the best effort service class may be acceptable for services in which there are no stringent constraints on delay (e.g. messaging, e-mail, file transfer, dial-up networking, etc.), it becomes a serious limitation for real time applications such as voice service, video telephony and videoconferencing. For public networks, it is necessary to define new mechanisms whereby the different types of service can be differentiated in order to protect real time traffic.

For this purpose, an architecture known as *Integrated Services* (Int-Serv) was developed which makes it possible to combine the robustness of a *connection-less* network with the real time service guarantees of a *connection-oriented* network.

This architecture, however, is not scalable and involves heavy management burdens in terms of signalling load. Thus, a new proposal called *Differentiated Services* (Diff-Serv) has been developed which is more suitable for application on wide area networks.

In the Int-Serv architecture, the quality of service is guaranteed for each individual data flow for which a request has been made. In this context, a flow is defined as a sequence of IP datagrams linked to the same user activity.

To reserve the resources needed in order to guarantee the required QoS for a given flow along the transmitter–receiver path, the RSVP

(*Resource reSerVation Protocol*) protocol is used. The main advantage of applying the Int-Serv architecture, and the RSVP resource reservation protocol in particular, is that an end-to-end QoS can be effectively guaranteed for each communication. Obviously, this is only possible if all of the networks involved in the communication between source and destination support RSVP. If this is not the case, there will be long spans where resource reservation signalling will have no effect.

In addition, the RSVP protocol enables each user to negotiate and if necessary change the QoS desired for a given service during a session.

Unfortunately, there is an inherent limitation in the way RSVP operates. The protocol is not scalable, and is efficient only for small LANs (*Local Area Networks*). As the size of the network increases, the amount of signalling traffic required grows significantly, calling for a large overhead in the routers which must store a state for each flow. These are major limitations in a UMTS context where, given the protocol's independence from routing algorithms, packets may be re-routed during a call onto a new path where the same QoS cannot be guaranteed.

With the *Diff-Serv* architecture, signalling message traffic is significantly reduced by grouping the various flows into service classes, which also makes for a drastic reduction in the various routers' computing load.

Each flow entering the domain in question is classified by a border router (BR) so that it can be handled consistently by the internal transit routers (TRs). Differentiated handling is accomplished by separating the traffic into queues according to the defined service classes. Flows are classified by marking an IP header DS field. The value contained in the DS field is then used by the network's internal routers for differentiated packet handling in relation to the QoS associated with each service class. In this way, it is no longer necessary to retain the soft-state of each flow in transit at each node, as was needed with the Int-Serv architecture.

An advantage of using Diff-Serv is that it can be applied to any network without limitations on size. In addition, it would be conceivable, thanks to preliminary agreements between operators, to define unique DS field values corresponding to the classes envisaged in the UMTS system in order to have uniform data handling (or in other words, the same guaranteed QoS) while roaming in other networks' domains. This freedom in associating DS field and QoS level is also a limitation, however, given that when flows cross networks outside those of the operators there is no guarantee that they will be handled in the same way.

In any case, the major limitation is connected with the fact that differentiated handling of grouped flows does not guarantee end-to-end QoS. In fact, the user has no feedback regarding whether the requested resources are available. Packets which were initially accepted and transmitted in the network may be discarded later if there are no available resources. Traffic is managed node by node on the basis of the different service classes, which correspond to different queues, without knowing beforehand whether resources are really available along the entire path. One way to get around this limitation would be to segment the packets before queuing them. In any case, it will be necessary to define intelligent algorithms for managing the various queues according the different QoS levels.

Another possibility would be to provide oversized links in order to guarantee that resources will be available to all classes. This approach is clearly not economical, and requires constant verification over time.

Finally, it would be possible to combine an Int-Serv architecture for small networks (e.g. Intranets and corporate networks) with a Diff-Serv architecture for large networks (e.g. core networks). In this case, it would still be necessary to define mechanisms which ensure that the two architectures can inter-work correctly. Specifically, common rules must be defined for translating Int-Serv service levels into Diff-Serv service levels.

References

[1] 3G TS 23.060 Draft GPRS Service description Stage 2.
[2] 3G TR 23.923 Combined GSM and Mobile IP mobility handling in UMTS IP CN, UMTS 23.20 Evolution of the GSM platform towards UMTS.
[3] TR 23.922 Architecture for All IP network.
[4] TS 23.107 Quality of Service, Concept and Architecture.

[1,2,4] © ETSI 1999. Further use, modification, redistribution is strictly prohibited. The above mentioned standard may be obtained from ETSI Publication Office, publications@etsi.fr, Tel.: +33 (0)4 94 42 41.

[3] TR 23.922 (Release 99) V1.0 is the property of ARIB, CWTS, ETSI, T1, TTA and TTC who own the copyright in them. They are subject to further modifications and are therefore provided to you 'as is' for information purpose only. Further use is strictly prohibited.

6

Opportunities for Satellites in Mobile Communications

Antonio Cavallaro and *Andrea Magliano*

This chapter presents the characteristics of the satellite systems which are most representative of the various families, illustrating the solutions now in operation as well as those being specified and developed for 2000–2005. We also have a look at competition and the current trends in the sector, describing the different types of customer involved.

To promote a clear understanding of the possibilities that are now available and of what we can expect the market to bring in coming years, any discussion of present-day and next-generation satellites should start by clearing away a few potential sources of confusion: satellite communications, like their ground-based cellular counterparts, have spawned a plethora of acronyms, abbreviations and conventional designations that are sometimes stretched far beyond their original meanings.

A case in point is the term UMTS. While continuing to take on new meanings in the terrestrial sector, the term has also been extended – as

S-UMTS, or Satellite UMTS – to satellite communications, where it has been used to mean different things at different times, viz.:

1   The satellite component used in the access network of third-generation terrestrial systems.

2   A satellite-based system which, independently of the terrestrial mobile network, can provide some or all of the capabilities envisaged for UMTS in terms of services.

3   A generic next-generation satellite system with entirely innovative technical features.

In some cases, these meanings overlap to a certain extent: the terms MSS (*Mobile Satellite Systems*), GMPCS (*Global Mobile Personal Communications via Satellite*), S-PCN (*Satellite Personal Communications Networks*) are used to denote systems, architectures and functions which are similar to each other and are often more or less deliberately confused with the term S-UMTS.

To clarify the real meaning and content of the term S-UMTS today, we must first provide a few explanations regarding the three definitions 1, 2 and 3 given above:

1   Though widely employed at ETSI, this use of the term is the most difficult to put into context. Basically, UMTS does not presently have a real satellite component which is in line with and has been brought to the same level of development now obtaining in the terrestrial sector. Moreover, it is still far from certain that such a component can exist in the future, even though groups such as the *Satellite UMTS Working Group* set up as part of ETSI's TC SES (Technical Committee Satellite Earth Station and Systems) are working in this direction.

2   This is the most likely interpretation if we consider the satellite systems that in the near future will be able to provide some of the ground-based UMTS system's capabilities. In addition to current systems for satellite mobile telephony (Iridium and competitors), this also includes future systems that will be able to provide chan-

nel capacity compatible with the decisions made for terrestrial systems.

3   Though this is the most debatable interpretation of the term S-UMTS, it should be mentioned because it is occasionally used to describe a new satellite system whose capabilities are innovative with respect to existing telecommunications networks.

6.1 Satellite systems for mobile telephony

6.1.1 Inmarsat

The satellite has played an important role in mobile telephony services for many years. It is in this context, for example, that we find the activities of (*INternational MARitime SATellite organization*), an inter-governmental organisation founded in 1979 for safety and communications at sea.

Over the years, the organisation has provided its users with a wide range of services through satellites based on well-established technologies. The satellites are in geostationary earth orbit to provide full coverage of the globe, excluding only the extreme polar regions.

In addition to the market segments which are strictly associated with maritime traffic, whether commercial, naval or recreational, Inmarsat decided some time ago to turn its attention to the aeronautical and terrestrial segments, promoting mobile telephony, data and fax services, Internet access, video communications and vehicle fleet management services. Partly in order to spotlight this new orientation, the organisation was re-named the *International Mobile Satellite Organisation* in 1994, though it is still known as Inmarsat for short.

At the end of 1998, there was a total of over 140 000 terminals using available services around the world, of which more than one third were portable terminals for terrestrial mobile applications. Inmarsat now offers high-level services for a wide range of customers: from journalists to broadcasters, from workers in remote sites to the armed forces and emergency services, and from naval and air fleet operators to passengers aboard aircraft and ships.

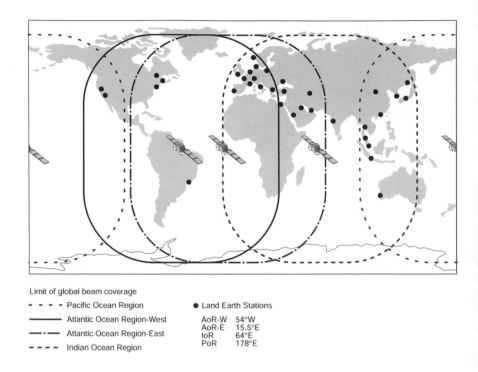

Limit of global beam coverage

- - - - Pacific Ocean Region   ● Land Earth Stations

———————— Atlantic Ocean Region-West   AoR-W   54°W
                                               AoR-E   15.5°E
—·—·— Atlantic Ocean Region-East   IoR   64°E
                                               PoR   178°E
- - - - Indian Ocean Region

Figure 6.1   Inmarsat: global coverage.

From the functional standpoint, the global communication system that Inmarsat provides today for its various services can be regarded as a satellite access network to the terrestrial telecommunications infrastructures. The system's global beam coverage is shown schematically in Figure 6.1.

As the arrangement of the four Ocean Regions used for satellite aiming indicates, the system was originally optimised for maritime services. The network's simple architectural design and the high reliability of its components make the system as a whole suitable for an extremely broad range of applications and ensure that it is open to future developments in services.

As shown in Figure 6.2, Inmarsat's customers in each ocean region can connect by means of a portable or vehicular satellite terminal known as an MES or *Mobile Earth Station* to one of the available *Land Earth*

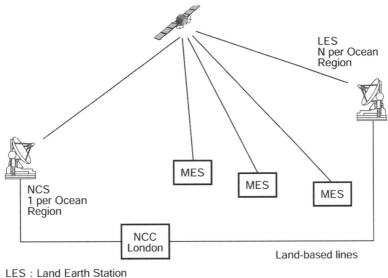

LES
N per Ocean
Region

MES

MES

MES

NCS
1 per Ocean
Region

NCC
London

Land-based lines

LES : Land Earth Station
MES : Mobile Earth Station
NCC : Network Control Centre
NCS : Network Control Station

Figure 6.2   Inmarsat: network architecture.

Stations (LESs), and thence to the land-based infrastructures in order to reach or be reached by the desired correspondent. Each terminal is identified by a unique telephone number, and calls in both directions can be made, as for terrestrial mobile telephony, simply by dialling the number of the desired correspondent.

According to current forecasts, Inmarsat will maintain its role as a key player for maritime communication and safety services. For other services, the process of privatisation which began in April 1999 could prove to be a decisive factor, over and above the actual size of the new markets. At the moment, the consortium – the world's leading inter-governmental organisation – is being transformed into a privately owned commercial enterprise, while headquarters are still located in London. Though ownership is still retained entirely by the consortium's members, who have agreed on their respective shares, it will be extended to include strategic partners. An initial public offering is expected to take place within the next few years.

6.1.2 The GMPCS systems

At the beginning of the 1990s, a large number of proposals were advanced for satellite-based systems belonging to the *Global Mobile Personal Communications* family. This trend, which drew considerable impetus from the sector's manufacturing industries, was to some extent an attempt to find alternatives to offset declining military investments in space technologies. Above all, however, it was a response to encouraging market forecasts concerning mobile telephony services.

The common objective of these proposals was to extend the coverage afforded by terrestrial cellular systems, which in that period appeared unlikely to spread beyond the more densely populated industrial areas. A distinctive characteristic of all of these projects was that they called for the deployment of a constellation consisting of a large number of satellites in a network capable of providing innovative performance features and world-wide coverage. A basic factor for all systems was the adoption of handheld terminals locked to the network at all times, even during stand-by, and thus capable of guaranteeing service with no need for the customer to carry out specific operations for aiming or accessing satellite radio channels.

To represent this family of systems, the following pages will provide a brief description of three projects: Iridium, Globalstar, and ICO. These projects have been promoted by three private partnerships that have polarised the sector's attention in recent years while continuing to compete on all fronts: from technical specifications to acquiring commercial licenses in the countries concerned, and from attracting financing to satellite construction and deployment. On the threshold of the new millennium, these partnerships, and the systems they have managed to set up, are faced with a mobile services market whose terms of reference have changed dramatically since the initiatives got under way. To cite a few of these changes, the coverage provided by terrestrial cellular systems, though still short of ubiquitous, is far greater than expected, standardised solutions such as GSM are enjoying undisputed success, call charges and the cost of terminals – already low enough to have fostered a mass market – continue to come down, and there is a clear need for services to evolve further.

Iridium

There can be no doubt that the Iridium system is innovative, complete and complex from the standpoint of its architecture and the technologies it adopts. The system was proposed by Motorola which, after presenting the project to the FCC (*Federal Communications Commission*) and obtaining a construction and deployment license in 1995, set up a company to promote the system together with partners from around the world, including China Great Wall Industry Corporation, Lockheed Martin, Raytheon, Korea Mobile Telecommunications, Sprint Corporation, Telecom Italia and Vebacom.

The network is based on a constellation of 66 operational satellites in low earth orbit (LEO) arranged in six orbital planes at an altitude of approximately 780 km from earth. The project derives its name from original plans to use 77 satellites, as 77 is the atomic number of the chemical element iridium.

Orbital planes are spaced at equal angles with 86.4° inclination, so that the satellites describe near-polar trajectories. The choice of a low earth orbit facilitates direct satellite radio channel access using compact handheld terminals, while the constellation's geometry guarantees complete coverage of the globe, polar regions included. The satellites complete one orbit in approximately 100 minutes. As a result, their footprint, or geographical coverage area, moves fairly quickly. This characteristic turns the whole concept of mobility upside down: it is no longer the handheld or vehicular terminal to move relative to the network, but the latter which moves, and at a speed that makes the users' movements negligible by comparison (Figure 6.3a,b).

The satellites, all of which are identical, are connected to user terminals with L-band radio channels; under line of sight conditions, they can reach the Gateway stations directly in the Ka-band. The satellites are also interconnected by means of fast channels called *Inter-Satellite Links* (ISLs), also in the Ka-band. Each satellite is provided with four ISLs which can be activated as needed. Two ISLs connect the satellite, respectively, to the one preceding it and the one following it on the same orbit, while the other two ISLs connect to the satellites immediately to the right and left on adjacent orbits.

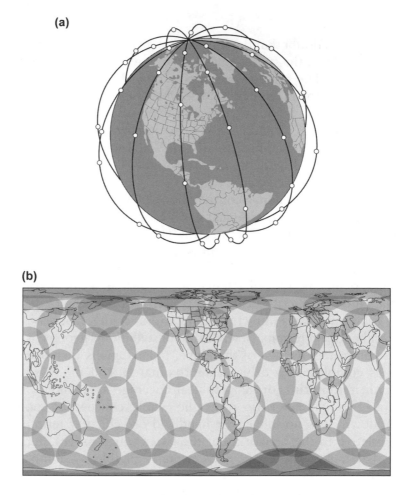

Figure 6.3   (a) Iridium: constellation; (b) Iridium: coverage.

The use of ISLs and on-board processing and switching capacity for digital signals to and from these connections are two innovative, and extremely important, features of the Iridium system. The satellite constellation provides a fully fledged telecommunications network in the sky, enabling two mobile customers anywhere in the world to connect to each other directly, without having to rely on the ground-based networks in any way. Customer mobility management

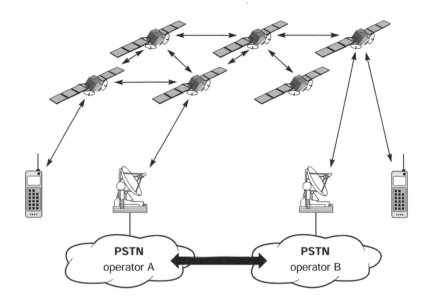

Figure 6.4   Iridium: network architecture.

is accomplished by means of a database system organised in two levels, with *Home Location Registers* (HLRs) and *Visitor Location Registers* (VLRs) as in GSM cellular networks (Figure 6.4).

The functions associated with mobility and, more generally, with satellite network control and management reside in the Gateway stations. When necessary, the latter also provide interfacing with the fixed and mobile terrestrial networks. As regards interconnection with other networks, it can be readily seen from the characteristics illustrated above that Iridium is also capable of operating with a single Gateway station, i.e. with only one interfacing point. Nevertheless, it was preferred to use a configuration with multiple Gateways (12 are envisaged) in order to increase redundancy, split up the processing load and, above all, to distribute traffic to the various regional areas. For traffic to and from Europe, for example, a Gateway has been set up near the Fucino basin in Italy.

As mentioned above, the system provides coverage for the entire globe from pole to pole. For user access, each satellite guarantees multiple spotbeam coverage, up to a maximum of a 48 spotbeams, or cells, distributed in a 4700 km diameter footprint. The spotbeams, obtained with fixed on-board antennas, sweep deterministically over the earth, following the satellites' motion. Complex hand-over procedures between spotbeams and satellites make it possible to compensate for the effects of this relative motion between the earth and beam coverage, guaranteeing that calls will not be interrupted.

Iridium provides its customers with a mobile telephony voice service. The system uses a 2.4/4.8 kbit/s audio coder, and each satellite can accept up to around one thousand voice circuits. QPSK digital modulation is used together with an FD-TDMA access technique. In addition to telephony, the project is also designed to provide data and fax services at a maximum bit rate of 2.4 kbit/s and a one-way fixed-to-mobile paging service.

Motorola and Kyocera are the only two manufacturers who hold a license to produce terminals capable of operating with the system. Two types of terminal are available: single-mode (via satellite only) or multi-mode (via satellite and via the terrestrial network, with various standards). Limitations common to all terminals are battery life and weight, which at around 400 grams is certainly higher than that of typical cellular telephones.

The first multiple launch of Iridium satellites took place in January 1997. Constellation deployment was completed during 1998, and in November of that year the partnership officially announced that it was ready for commercial service.

In Motorola's original plans, Iridium was to attract not only a vertical market of niche customers (military, emergency services, workers in remote areas), but was also – and chiefly – designed for a rich and sizable population of businessmen, frequent travellers who needed a universal mobile phone with world-wide range. With the passage of time, this target has been revised in favour of more traditional market segments.

Globalstar

Supported by an international group whose partners include Loral Space and Communications, Qualcomm, AirTouch, Alcatel, Alenia, Daimler-Benz Aerospace, France Telecom, Finmeccanica, and Vodafone, this system was designed to provide mobile telephony services in

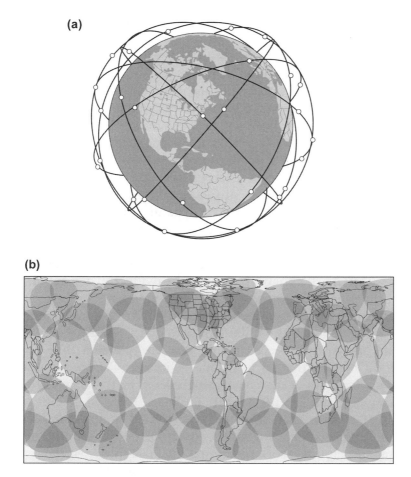

Figure 6.5   (a) Constellation; (b) coverage.

a band on both sides of the equator between 69° of north and south latitude (Figure 6.5a,b).

Though coverage is thus not global in the strict sense of the term, all populated areas of the earth of potential commercial interest are served.

The 48-satellite constellation is based on eight LEO circular orbits at an altitude of 1410 km. Inclination is 52° and the orbital period is 114 minutes.

The satellites provide geographical coverage by projecting 16 spot-beams with a diameter of approximately 2200 km. They are connected to user terminals by means of S-band radio channels shared through a code division multiple access technique (CDMA); connection to the Gateways is accomplished using C-band channels and the FDMA-FDM frequency division access technique.

There are no direct links (ISLs) between the satellites, and on-board systems are not capable of performing signal processing and switching functions. This is a major difference between the system and Iridium. Though it results in greater simplicity, it also makes it necessary to use a sizable number of Gateway stations: 50 in the current deployment plan. For the customers dynamically present in its service area, each satellite sets up access connections to the terrestrial infrastructures, and can consequently operate correctly only when at least one Gateway is in its line of sight. Communication between two mobile customers cannot be established directly, terminating on board the satellite, because signal switching must take place on the ground in the Gateway network. From the architectural standpoint, there are a number of similarities with the GSM network, as the satellite can be regarded as performing the role of a Base Station, and the Gateway that of a Mobile Switching Centre. Here again, customer mobility is managed using classic methods, with a two-level database (HLR-VLR) (Figure 6.6).

With its transparent satellites, the selected architecture concentrates all of the system's complex functions on the ground, making maintenance and network upgrading much simpler. On the basis of original pricing announcements, costs for the end user should be competitive, especially as regards the cost of the terminal.

Because of the particular nature of the choices made in its design, there can be no doubt that Globalstar is best regarded as an extension of the terrestrial networks, and is intended to play a highly complementary role. Rather than targeting frequent travellers, its market will chiefly be found among developing countries which do not yet have adequate ground-based infrastructures, companies with operations in remote areas, and niche customers whose needs cannot be satisfied by the existing terrestrial networks.

Globalstar will provide its customers with mobile telephony through an audio coder capable of adapting the output data stream

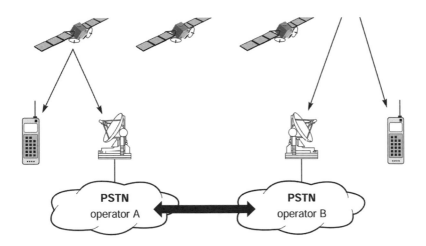

Figure 6.6   Globalstar: network architecture.

rate (from 2.4 to 9.6 kbit/s) to the user's voice activity. This optimises the system's capacity, which can be estimated at around 2000 voice circuits per satellite. Messaging, fax and data services will also be provided at bit rates up to 9.6 kbit/s.

Users will be able to choose between single- and multi-mode hand-held terminals, which will be produced by Qualcomm, Ericsson and Telital.

By August 1999, 36 satellites had been launched, a number sufficient to begin service, though in some cases it will be necessary to accept a reduction in the number of satellites which are visible at any one time. On the same date, it was announced that the constellation would be completed by the end of the year, and that it would be possible to start commercial service on a regional basis as the Gateway stations were deployed.

ICO

This system is the brainchild of the Inmarsat partnership, which originally promoted a series of preliminary technical and market studies for the project. Development and deployment is now handled by ICO Global Communications, a publicly traded company established for the purpose in 1995.

Investors include Inmarsat Ltd, Beijing Marine Comm. & Navigation Corporation, British Telecom, Comsat Corporation, DeTeMo-
bil, Hughes Electronics Corporation, Korea Telecom, OTE, Satellite
Phone Japan Ltd, TRW and VSNL.

Of the three systems we have considered, there can be no doubt that
ICO is the one based on the least innovative, and hence best consolidated, technological solutions (Figure 6.7a,b).

(a)

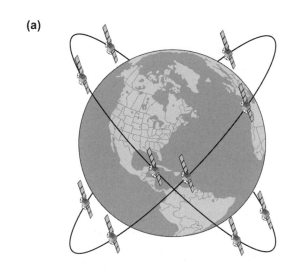

(b)

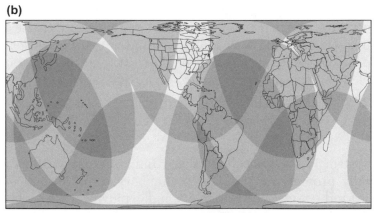

Figure 6.7   (a) ICO: constellation; (b) ICO: coverage.

The constellation consists of ten satellites on two circular MEOs (*Medium Earth Orbits*) at 10335 km above the earth's surface. Orbital planes are inclined at 45° to the equator. The satellites' orbital period is approximately 359 minutes. There are no ISL links.

Terminal access to S-band satellite radio channels is based on TDMA time division techniques, whereas connection with the Gateway stations, which for this system are called *Satellite Access Nodes* or SANs, uses the FDMA-FDD frequency division technique.

This orbital configuration was chosen after extensive studies, conducted initially by Inmarsat and Hughes engineers. The objective of these studies was to achieve the best trade-off between economical implementation and simple system management. In fact, opting for MEO orbit made it possible to:

• Reduce the number of satellites required for global coverage.

• Permit a larger elevation angle and ensure that several satellites will be in view under favourable conditions.

• Reduce the satellites' speed relative to the earth, thus making hand-over management simpler than in LEO systems.

On-board systems are not provided with signal processing and switching capabilities, but permit user terminals to access the ground infrastructure consisting of 12 SAN stations located around the globe. Customer mobility databases are located at the SANs. These stations are extensively interconnected through the dedicated ICONET ground network, which is needed to connect mobile users served by satellites in the line of sight of different SANS, as well as to connect the ICO system to the outside world, minimising reliance on the terrestrial networks.

ICO will provide its customers with mobile telephony through a 4.8 kbit/s audio coder. System capacity can be estimated at approximately 4500 circuits per satellite. Messaging, fax and data services will also be provided.

Here as in the other systems, multistandard terminals are envisaged. As early as 1998, specific contracts to supply these terminals had already been drawn up with a variety of manufacturers, including Hughes Network System, Matsushita Communication-Panasonic,

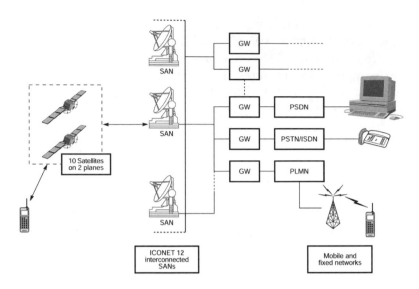

Figure 6.8   ICO: network architecture.

Mitsubishi, NEC Corp. and Samsung. Service startup is scheduled to take place before the end of 2001 (Figure 6.8).

6.2 The Super-GEO systems

The term Super-Geo is used to designate all satellite systems which will provide mobile telephony services on a regional or continental basis using geostationary satellites. In past years, given certain limitations of geostationary satellites (such as their high latency and inability to provide coverage for the earth's entire surface), this approach was thought to have been superseded by the LEO and MEO schemes discussed above in connection with GMPCS systems, which are better in terms of latency (a few dozen milliseconds, as compared with 260 ms) and global coverage. On the other hand, the cost of deploying constellations and, in general, the complexity of implementing a satellite network are considerably higher, a fact which goes far to lessen the attractions of low earth orbiting systems. For these and other reasons, solutions based on geostationary satellites are, paradoxically, healthier than ever before, just when LEO and MEO systems are coming onto the market.

ACeS, Thuraya, EAST, Agrani are a few of the Super-GEO systems which, though not nearly as well known as the GMPCSs, are in a good position to compete against them in their own particular areas.

To represent this family of solutions, a brief description of the ACeS and Thuraya systems will be given below.

ACeS (Asia Cellular Satellite)

The ACeS project will use two GEO satellites (GARUDA-1 and 2) produced by Lockheed Martin. With satellite launching and the beginning of commercial service scheduled for late 1999, ACeS will be the first system of its kind to operate in the Asia-Pacific region. Coverage will be concentrated on the world's most densely populated areas, while two thirds of the earth's population (Japan, Korea, China, India, and all of South-East Asia, including Indonesia) will be within range of the ACeS signal.

Voice and data services will be provided together with satellite-GSM and satellite-AMPS roaming. The first Service Providers have been set up for India, Bangladesh, Pakistan, Indonesia, Thailand, the Philippines, Taiwan, Japan and Sri Lanka. In addition to Lockheed Martin, which holds 30 percent, the major stockholders are Ericsson (supplier of the terminals) and several of South-East Asia's national operating companies, including PT Pacifik Satelit Nusantara for Indonesia, Jasmine International Overseas Co. for Thailand, and Philippines Long Distance Telephone Company. The satellite will be launched on a Proton rocket, and will be managed by a control station in Indonesia. Traffic charges are currently expected to be in the neighbourhood of US $1 per minute.

Though deployment is lagging behind announced schedules, the system has now attracted financing for the entire 900 million dollars needed for set-up, and could be the world's first super-GEO system to enter service (if we do not include the United States' long-established AMSC in this category, as its characteristics are markedly different).

Thuraya

Thuraya will probably be one of the first super-GEO systems to enter the mobile telephony market. It is of considerable interest for Italy, as coverage extends to all of Europe (except for Scandinavia and Russia), as well as north and central Africa, the Middle East and the former southern Soviet republics as far as India.

Two orbital positions at 44°E and 28.5°E have been reserved for the system. Initially, Thuraya will have a single geostationary satellite constructed by Hughes Space & Communication. An interesting feature of this satellite will be its use of a 17 m diameter unfolding antenna, which is needed to achieve sufficient reception gain to use handheld terminals.

The ground segment will feature Gateway stations supplied by Ericsson, with independently operating regional Gateways and a main Gateway located in the United Arab Emirates, whose telecommunications operator is the system's chief backer.

The system's objective is to provide mobile telephony, fax and data services via handheld and transportable (i.e. vehicular) dual-mode GSM and satellite terminals. The terminals will be made by Hughes and Ascom and are expected to be sold at prices around US $600. Traffic charges for the end user are likely to be in the US $0.50–1.50 per minute range.

At the moment, the group which supports Thuraya is made up of six strategic partners and 12 national operators. In July 1999, the group had raised the entire 1100 million dollars needed to deploy the system. Service is scheduled to begin in the autumn of 2000, and with the coverage indicated above should be available in around 100 countries.

6.3 Third-generation mobile telephony: the distinctive features of satellite-based solutions

The groups that are now working to establish third-generation terrestrial cellular systems are attempting to follow an approach which has proved fruitful in the past: agreeing on a common standard at the continental or, even better, the world-wide level could produce enormous economic returns for all operators, manufacturers and service providers.

By contrast, satellite communications are inherently competitive, even from the strictly technical standpoint, and situations where different systems are able to coexist are few in number and unstable by nature.

Simplifying somewhat, we can say that whereas technical convergence and commercial competition is possible and often desirable in terrestrial circles, competition is far more deeply engrained in the world of satellites, and agreements of any kind between groups with divergent interests are more difficult to reach.

Consider, for example, the competition for frequency allocations in the various bands, the struggle to obtain adequate orbital positions in the geostationary orbit, and the contracts for exclusive provisioning rights. All this is due to technical and technological factors, as well as to economic considerations. Thus, a satellite telecommunications network has several distinctive features which set it apart from a similar terrestrial network, and illustrate the differences between the two.

1   Different *network deployment modes*. A satellite network is typically deployed in what can be called an on–off mode: in other words, the network is in a certain sense deployed all at once. By comparison with a terrestrial network, this has advantages (greater speed, complete service in the entire coverage area) and disadvantages (investments are hard to spread out over time, planning errors are difficult to correct).

2   Higher *up-front costs*. The massive initial investments needed to set up a satellite network, particularly if large constellations are used, means that these projects are most likely to be taken on by partnerships, joint ventures and agreements between manufacturing industries. Satellite television broadcasting in Europe provides an interesting example in another setting: two partnerships split much of the market between them, and the fact that the same standard is used is by no means the result of an agreement between the two.

3   *World-wide market*. Even more than is the case for terrestrial networks, the complexity of the struggle to gain the world satel-

lite market makes the economic, commercial and regulatory (i.e. agreements and licenses) aspects paradoxically more important than the strictly technical considerations – despite the enormous technical and technological challenges posed by the new satellite system's many innovative characteristics.

To sum up, we can say that satellite communications are marked by stiff competition, little inclination towards broad-based technical agreements, and slight interest in establishing standards through participation in more or less official work groups and standards-writing organisations in general. These characteristics, though shared to a certain extent by the 'terrestrial' world, are carried to extremes in that of satellite communications.

The two environments ('terrestrial mobile' and 'satellite mobile'), though traditionally separate, are nevertheless destined to have interesting points of contact in the future (Figures 6.9 and 6.10).[1]

6.4 Standardisation groups: the current situation

The world of satellite telecommunications has always been characterised by several aspects which differ markedly from the 'terrestrial' world, showing it on the whole to be far more prone to conflict:

- *Few large operators*: even among the manufacturing industries themselves, the market is split up between a few giants.

- *Military origins*: many of the major industries working in the field today retain close links, including economic ones, with military circles, with everything this implies in terms of restricted access to information (security) and a tendency to prefer strong-arm tactics, mergers and buy-outs to agreements between companies on an equal footing.

(1) In the two figures, the size of the oval indicates the range of services provided and their geographical extension, rather than market share. The relative standing of each technology and operator in the wireless market can thus be seen from a typically satellite-oriented perspective.

- *Predominance of partnerships*: partnerships such as Inmarsat, Intelsat and Eutelsat, though operating in profoundly different settings, have been a constant in the history of satellite telecommunications over the last thirty years. This exception to the

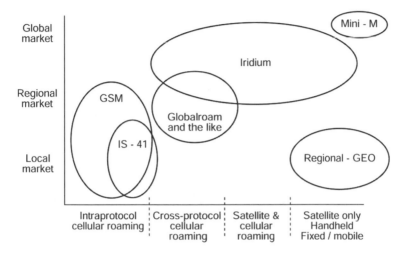

Figure 6.9 Competition in satellite mobile communications (today).

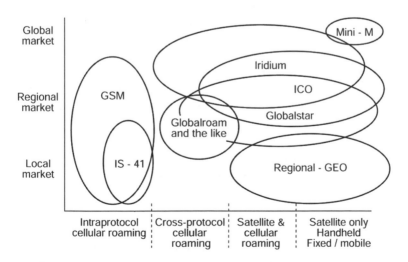

Figure 6.10 Competition in satellite mobile communications (after 2000).

conflict-prone attitude mentioned above is chiefly due to the fact that the participants in these partnerships were and are essentially representatives of governments rather than of industries.

Though satellite communications live in a world which is far less 'co-operative' than that of their terrestrial counterparts, it is interesting to follow the efforts which all – or almost all – of the major operators in the satellite sector have recently been making in order to determine what form the satellite component of third-generation mobile telephony systems will take.

6.4.1 ETSI TC-SES (Satellite Earth Stations and Systems)

The ETSI Technical Committee SES (Satellite Earth Stations and Systems) has recently set up a roundtable addressing S-UMTS which has drawn participants from those manufacturing industries that have traditionally refused to take part in initiatives of this type, preferring to impose (or to attempt to impose) unilateral solutions designed to put the other contenders out of the running at the outset.

As will be recalled, ETSI's TC SES is responsible for everything concerning satellite communications (including mobile communications and broadcasting), earth stations and protocols used in satellite communications.

The Satellite UMTS working group held five meetings in the 1998–1999 period which were attended by many of the major satellite and terrestrial cellular operators, including Inmarsat, ICO, Nera, Nokia, Siemens, Hughes, ESA, Telespazio, Ericsson and others.

The main topics of interest addressed by the working group are listed below:

- Specifying requirements for the UMTS satellite component in the following areas: services that can be provided, SIMs, Terminal families, Network interfaces.

- At the architectural level, identifying the crucial differences between typical terrestrial and satellite architectures and, at the interface level in particular, examining terrestrial interfaces that could be suitable for satellite use.

- Establishing liaisons with the major standards-writing organisations and interest groups who are active in the sector, starting from SMG12, but also including SMG2, T1P1, UMTS Forum, GSM MoU, ACTS and ITU.

- Determining type-approval requirements for single, dual and multimode satellite terminals.

Further meetings of the S-UMTS WG are scheduled for the near future. In view of the increased participation by satellite manufacturers mentioned above, it is likely that they will be of a certain interest for terrestrial operators as well.

Thus, while on the one hand the meaning associated with the term UMTS in terrestrial circles is becoming increasing clear thanks to the collective decisions taken on a planned basis by the standardisation bodies, the satellite component has yet to take on firm outlines, a situation which – as we have discussed above – is largely due to the satellite operators.

As a result, keen interest now attaches to the efforts which these operators are making to establish a common ground for specifying the system, or at least a portion of it, be it only the radio interface (to take one of the UMTS elements for which attempts to develop standards have generated the fiercest disputes between operators and manufacturers).

In recent months, the TC SES has also set up several new working groups addressing satellite topics. The most important of these topics and groups are listed in Table 6.1.

Table 6.1   Currently active ETSI WGs

Topic	WG starting date
Satellite UMTS	April 1998
Multimedia Satellite Broadband	April 1998
Space Standardisation	February 1998
S-PCN GEO	November 1997
Little LEOs below 1 GHz	November 1997

6.5 Stratospheric platforms: an alternative?

The idea of providing communications services to major cities by means of signal repeaters located on high towers or on dirigibles, balloons or even on very high altitude aeroplanes is by no means new. Nevertheless, these often fanciful proposals were entirely within the realm of science fiction and, though exercising considerable appeal for the imagination, were never anything more than hypothetical and were never taken into serious consideration either by the aeronautical and satellite industry, or by telecommunications in the broad sense of the term.

Despite their obvious potential from the standpoint of telecommunications *per se* (cost, simplicity, traffic capacity), the technological obstacles to implementing such schemes have always been all too clear.

Today, the situation has to some extent changed. A number of projects, some still in the embryonic stage, some more advanced, are now planning to deploy platforms operating in the stratosphere at altitudes of more than 20 km from earth.

We are thus beginning to hear of HAAPs (*High Altitude Aeronautical Platforms*), a term which covers several radically different systems which taken together are noteworthy from the point of view of communications and system characteristics.

Broadly speaking, HAAPs can be divided into two categories. The so-called *stratospheric platforms* call for placing a communications payload beneath a balloon similar to a dirigible. This balloon would then be put in the stratosphere above a metropolitan area. Obviously, many dozens of these balloons would be needed to cover all of the areas of interest effectively. The most important of these systems proposing to use balloons is Skystation. The second category is that of high-altitude aircraft, or in other words of special aeroplanes flying in the stratosphere and operating as high altitude radio stations. The HALO (*High Altitude Long Operation*) system proposed by Angel Technologies falls into this second category.

These two systems, Skystation and HALO, are currently the only ones to have significant technical content and the backing of an industrial group.

However, there are still a number of knotty technological problems to be solved. The main obstacles, though doubtless not the only ones, are those associated with stabilisation – in the case of stratospheric balloons – and with solar battery power supply in the case of high-altitude aircraft. In the first case, ion propulsion, a technology which is beginning to show promise for space missions, is not sufficiently mature and does not at the moment appear to be capable of coping with an environment as difficult as the middle atmosphere. Even in the second case, aerodynamic considerations give rise to doubts concerning these systems' feasibility.

Though these doubts persist, stratospheric station telecommunications systems should be borne in mind for the medium-long term, although it is unlikely that they can be implemented sooner. In any case, their 'competitiveness' in terms of cost, end user data rate and ease of deployment makes them a future possibility which should be carefully assessed if and when the technological limitations which now stand in their way are overcome. In particular, these system's great capacity per unit of terrestrial surface area makes them interesting candidates for providing UMTS services in metropolitan areas, i.e. precisely in those areas where they are most attractive and where the limited bandwidth around 2 GHz makes it difficult to supply high bit rate services to a large number of users in a geographically restricted area using traditional methods such as terrestrial cellular systems and satellites.

References

[1] CALCUTT D., TETLEY L., 1994, *Satellite Communications: Principles and Applications*, E. Arnold.
[2] GARDINER J., WEST B., 1995, *Personal Communications Systems and Technologies*, Artech House.
[3] HATLELID J. E., CASEY L., 1993, *The Iridium System: Personal Communications Anytime, Anyplace*, Proc. 3° IMSC, International Mobile Satellite Conference, Pasadena.
[4] OHMORI S., WAKANA H., KAWASE S., 1998, *Mobile Satellite Communications*, Artech House.

7

Terminals and Applications

Giorgio Castelli

7.1 The evolution of mobility services

The signs of market evolution and the many and varied corporate agreements and movements clearly indicate that an increasingly important share of the innovations in the mobile systems and services sector arise as a result of evolving mobile data services.

This growth trend in services is apparent in terms both of differentiation and of quantitative impact. In this sense, mobile data services will in the near future be one of the high-growth and, presumably, high-profit sectors for service providers and operators, even if the overall volume of these services at the moment is insignificant by comparison with that of voice services. In a market setting of this kind, the main drivers for innovation spring from the close synergies between:

- The size and spread of mobile services.
- The parallel development of Information Technology services and applications.

These synergies are fostered by the rapid advances made in mobile radio terminals and development tools, which in turn encourage continuing enhancement of services and applications.

In this context, the opportunities provided by the introduction of UMTS third-generation mobile networks will be instrumental in ensuring fuller use of the services made available by greater mobile transmission bandwidth.

Consequently, the technical and market requirements for data applications will call for radical changes in how the radio resource (packet) is used and in overall network architecture.

In the fast-paced and hypercompetitive mobile radio market, the operator must be able to respond with solid strategic abilities in terms of innovation and customer care. Consequently, the development of application and systems-related solutions for the mobile environment will be ever more closely linked to the user's needs. Sky-rocketing growth in the number of mobile customers is accompanied by extremely rapid terminal turnover as customers replace their older units with new ones, thus creating a prime opportunity for introducing new services which are also based on innovative terminal characteristics. Even today, as regards access to value-added services, a distinction is arising between 'voice centric' terminals or 'smart phones' and 'information centric' terminals, or 'communicators'. This trend will gain strength in the third generation, with multimedia terminals for video services and advanced navigation.

For the GSM system, new application environments have recently become available for providing value-added services (VAS) for GSM mobility. Examples include the SIM (*Subscriber Identity Module*) Application Toolkit, which makes it possible to personalise the SIM with new applications, WAP (*Wireless Application Protocol*), a protocol stack developed in order to access Internet services via cellular telephone, and MexE (*Mobile Execution Environment*), an ETSI standard founded on WAP and the Java programming language.

This changes the outlook for developing and supporting data applications, or integrated data and voice applications, for both the horizontal market and for business customers.

The WAP services provide interactive access to Internet applications and information using only a GSM cellular phone. This is possible thanks to the introduction of a micro-browser in the cellular phone, so that value-added application services can be provided to businesses and to individual customers. These services, which range from economic or sports news to mobile banking and mobile commerce, are based on Internet protocols optimised for cellular phones' limited bandwidth and display capabilities. Connection takes place with data transmission (which in the future will be packet-based) to the mobile network, where a WAP server converts protocols between the GSM mobile environment and that of the IP application servers (Figure 7.1).

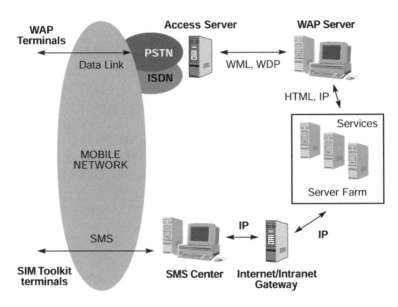

Figure 7.1 Value-added service architecture for the GSM network.

The services based on SIM Application Toolkit take advantage of the increasing processing and memory storage capacity offered by SIM cards (which currently provide 16 kbytes, of which 8 are dedicated to three or four applications available for the customer), and are an important means of reinforcing customer retention, given that the customer may perceive the SIM as a feature which sets the operator apart from the rest. Value-added services such as home banking, e-commerce and location-based applications, moreover, can be customised through an over-the-air download mechanism whereby the applications are downloaded via the GSM network, and are an approach which complements WAP. Connection takes place by sending and receiving menu-prompted short messages to the mobile network, which forwards them to application servers.

The scenario is made even more complex by new GSM phase 2 + bearer services such as GPRS (*General Packet Radio Services*). Through these services, users will have a good preview of what they will be able to expect when today's GSM mobile network evolves towards broadband services suitable for the IP world. The gradual spread of GPRS will provide users with Internet–Intranet type communication capabilities (packetised data transmission at medium/high bit rates) and with advanced terminals with the kinds of display and keypad needed for these services.

As a result, there is a need to establish a 'technological mix' for introducing each new service, considering the various alternatives for terminals, bearer services, support infrastructures (SIM Application Toolkit, WAP) and types of evolution towards UMTS.

As regards the application solutions, which will be made available by the UMTS third-generation mobile network, the main trends will be guided by the development of new terminals (e.g. provided with high-definition colour displays or integrated TV cameras) and by the availability of packet switched broadband access.

The biggest problem associated with implementing UMTS terminals, in any case, consists in being able to guarantee global roaming. This entails supporting the various CDMA techniques that will be used, as well as maintaining compatibility with second-generation mobile radio systems. At the moment, even the manufacturers themselves are by no means certain that this can be achieved, and expect that various classes of terminal will be available in the future. In their view, these classes will be distinguished on the basis of the second-generation techniques and networks they support, as well as by the services they provide. As the divisions between cellular phones and palm-top computers are expected to become increasingly indistinct, manufacturers are now looking into the possibility of tackling the problem with software rather than hardware solutions, and a few prototypes have already been produced.

7.2 Mobile terminal evolution and market prospects

Burgeoning growth in the number of mobile customers is accompanied by extremely rapid terminal turnover, as around 70 percent of customers own a terminal which is less than one year old. There is thus an excellent opportunity for introducing new services based on the terminal's features.

Yearly sales of cellular phones in Europe will pass the 100 million mark from the year 2002 onwards, while growth trends are expected to show substantially the same patterns on the Italian market (with a target of 20 million terminals for 2002) and the world market (over 500 million in the same year) (Figure 7.2).

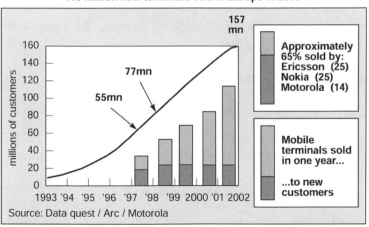

VOICE TERMINAL GROWTH TRENDS

An enormous market:
110 Million new terminals sold in Europe in 2002

Burgeoning growth in the number of mobile customers accompanied by extremely rapid terminal turnover (70% of users own a terminal which is less than a year old)

Figure 7.2 Sales of voice terminals.

7.2.1 Second-generation terminals

Another issue to be borne in mind is that of the capabilities that have already been introduced by the major manufacturers, and which will characterise the new mobile terminals to an ever-increasing extent.

The differentiation in the types of mobile terminal, permitting more advanced operations and services, will take place at the point of convergence between the world of classic cellular phones and that of Personal Computers (PCs). Terminals will thus be classed in one of two families, according to which of the two forebears they are closest to:

- Information centric: communicator.

- Voice centric: smartphone.

The major companies in the ICT (Information and Communication Technology) sector are already working along these lines, proposing the basic Operating System as a fundamental characteristic of the terminal, which they refer to specifically as an 'information appliance':

- Symbian (supported by Nokia, Ericsson, Motorola) with its EPOC Operating System.

- Microsoft with Windows CE (Consumer Electronics).

- 3Com with the Palm Operating System.

Even in the future, however, these devices will account for a limited percentage of the enormous number of classic voice terminals.

7.2.2 Advanced third-generation terminals

The diversification of mobile terminals will be even more marked for the UMTS network. Given the emphasis on the multimedia services this network will provide, terminal manufacturers are working to ensure that these services are as user-friendly as possible.

The figures on the following pages show a series of WB-CDMA UMTS mobile terminals developed for the tests which the Japanese network operator NTT DoCoMo is conducting with a view to putting third-generation IMT2000 services on the market. At the moment, these terminals are prototypes, and it is not certain that they will be mass produced in the form shown. However, they give a good idea of what will be available in the future.

There will also be a market for terminals similar to today's cellular phones, which will be used essentially for voice communication but may also have messaging capabilities. One of the main objectives is that these third-generation mobile phones maintain the major attractions of their present-day confreres, or in other words, handiness and light weight. This goal can be summed up in the 'four 100s rule':

- 100 g weight;

- 100 cm^3 volume;

- 100 h stand-by;

- 100 MIPS (Mega Instructions Per Second) (Figure 7.3).

Multimedia terminals will be the real novelty for UMTS (though it seems that the GSM-GPRS network will provide a foretaste of their capabilities), and will be based essentially on integrating advanced technologies for the colour display, audio and video recording using micro-TV cameras (Figure 7.4).

The figure below shows European market forecasts for the different types of multimedia service (Figure 7.5).

Naturally, there will also be data transmission terminals which will be integrated in normal PCs through the use of PCMCIA (*Personal Computer Memory Card International Association*) cards.

Given the high rate of growth for the 'mobile Internet' phenomenon, forecasts for the sustained development of terminals for UMTS Web access are based on an evolutionary pathway which starts from WAP and, thanks to GPRS, takes a course which is already clearly outlined today (Figure 7.6).

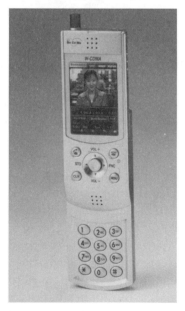

Figure 7.3 UMTS mobile terminal prototypes.

Figure 7.4   UMTS mobile terminal prototypes for multimedia services.

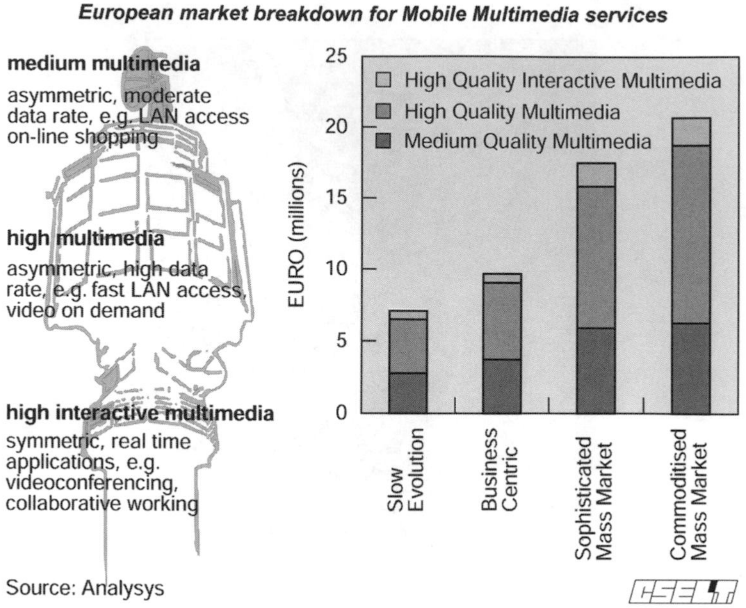

Figure 7.5 European market breakdown for mobile multimedia services.

The main problem associated with implementing third-generation terminals, however, consists in being able to guarantee global roaming, one of the stated objectives of the UMTS system. The situation is complicated by the fact that at the world-wide level the third-generation solution will consist of a family of systems, each with its own radio features. At the same time, it will be necessary to guarantee backward compatibility with second-generation systems, as discussed in the preceding chapters.

The problem is thus more complex than that, for example, of developing dual-band telephones, though even the latter called for enormous exertions on the part of manufacturing industries. Forecasts indicate that the market will probably be characterised by terminals which will also be differentiated on the basis of the radio interfaces they support and their compatibility with second-generation systems. Thus, not all terminals will be capable of supporting global roaming.

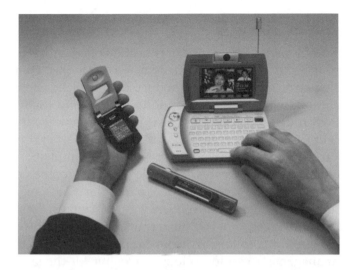

Figure 7.6 UMTS mobile terminal prototypes for Internet services.

To solve this problem, a number of manufacturers have proposed a complementary solution. Given the difficulties in achieving convergence towards a single world-wide standard, they suggest that the largest possible portion of the standards be harmonised, and that the remaining differences be implemented in software rather than directly in hardware; subsequently, they believe, radio terminals should be developed whose radio part is controlled entirely by hardware. Though not all manufacturers agree with this approach, it should bring costs down and help reduce the problem's complexity.

This solution, which is known as software-radio (a topic that will be discussed in detail in Chapter 9), is in fact a new frontier in the development of mobile terminals. At the moment, several prototype solutions are already available which are able to implement part of the radio interface. The greatest obstacle to be overcome consists of these terminal's high consumption, particularly as regards the modules for controlling high-frequency radio signals. In any case, the expected convergence of mobile terminals with handheld computers (PDAs) could make this problem insignificant, given that radio interface management would no longer account for the largest share of power and processing capacity.

7.3 UMTS services

One of the main goals of UMTS is to make it possible to provide a wide range of voice, data and multimedia services in an extremely competitive and fast-moving environment. This should pave the way to a scenario in which operators could decide to specialise in roles other than those they have traditionally covered, e.g. acting as service and content providers, and, at the same time, where other companies who have always been excluded from the mobile radio market could gain a foothold on it.

The objective will thus not only be that of offering an increasingly wide range of ever more innovative services, but also that of guaranteeing an integrated, personalised and homogeneous environment for the user independently of the type of terminal or network used to access the service. In particular, to use the terminology employed in the world of telecommunications, it will be necessary to guarantee what is commonly referred to as *terminal mobility*, *user mobility*, and *service mobility*.

The term *terminal mobility* means the ability to use the services that have been subscribed to regardless of the specific access network involved. *User mobility*, on the other hand, means the ability to use the services that have been subscribed to regardless of the type, make and model of terminal employed. Finally, *service mobility* refers to the user's ability to access personalised services independently of the terminal and network employed.

For these purposes, UMTS will use the smart card technology developed for GSM and will offer what is referred to as a Virtual Home Environment (VHE), or in other words, an environment in which the user will benefit from the same services with the same interface he originally chose, regardless of the specific location or terminal from which the service is requested

The following sections will provide a more detailed analysis of the VHE concept, its impact on UMTS service creation, and the main classes of third generation services. These services will be chiefly characterised by their ability to support multimedia content and the large bandwidth at their disposal, at least by comparison with today's second-generation mobile radio systems. Obviously, dividing the services into hard and fast categories is not possible, as the fact that varying technologies and market segments are integrated means that the applications tend by their very nature to cut across several classes. Consequently, we will attempt to characterise the services on the basis of their major features or of the customers to whom they are chiefly addressed.

7.3.1 Virtual home environment

As the name implies, the *Virtual Home Environment* is a virtual environment which UMTS establishes in order to guarantee that the user can access subscribed services using the same user-selected methods from any terminal or network. The user thus has the impression of being on his or her own home network even while roaming on that of another operator, with clear advantages for all of the parties involved (e.g. operators, customers and service providers).

VHE must enable terminals to negotiate functions with the access network, and thus must also make it possible to download the software which provides functions and services from the home network, seamlessly and with complete security for the user.

In many people's view, seamless VHE mechanisms will be the major factor in achieving a mass market for UMTS services.

However, VHE does not limit itself to established methods and restrictions for using services, but also determines how they are developed. Accordingly, it includes a service creation environment to support the rapid growth of UMTS services. The basic concept is to make service creation, introduction on the network and portability as easy as possible by specifying a single communication interface with the UMTS system and a set of support tools for design and testing. The aim is thus to make service creation independent of the specific network, type of access or terminal that will then be used.

In this way, a standard development platform is established which enables independent groups to create or distribute services to their customers via UMTS, in some cases adapting information content which is already available through other types of access.

The flexibility introduced by VHE will not only increase the number of services that can be offered, but will also make it possible to characterise them on the basis of the market segment they target and, in the most optimistic view, on the basis of the individual customer's needs. Thus, it will not only be possible to use the desired service anytime and anywhere, but even to adapt to varying access conditions such as the type of terminal or network, or to specific user requirements.

7.3.2 Multimedia services

Multimedia services are one of the most attractive, and indeed fascinating, features of UMTS. Available bandwidth will make it possible to develop a wide range of services that can be used directly from the customer's mobile terminal. There is a multitude of conceivable applications, which could be supported by terminals with widely differing characteristics. In addition to video-communication or video-streaming applications, services such as messaging and navigation that already exist today, and which will be discussed in the following sections, could evolve and be integrated. The main applications based on transmitting video content could be as follows:

- Video conferencing applications, thanks to the use of video-phones equipped with miniature video cameras. In addition to making audio and video calls, it will be possible for example to make an audio-video recording of a meeting and transmit it to co-workers, or to provide video support for remote training and collaborative working.

- Video-streaming applications, where films can be received by request directly on the cellular phone. The service will be char-acterised by the nature of the material involved, which could range from music video-clips, to sports films, movie trailers, etc.

- On-line video sales catalogues. For example, real estate agents could provide on-line access to their property listings so that potential buyers would not have to visit houses on sale personally.

- Telemedicine applications. It would be conceivable, for instance, to transmit accident victims' X-rays or photographs directly at the site of the accident.

7.3.3 Access to Internet–Intranet services

Mobile telephony and Internet access are the two technologies that have had the greatest impact on the market in recent years. Their growth, in terms of world-wide users, has been truly explosive. It is thus interesting to follow the course they are taking towards integra-tion, which has already started on the GSM network with the intro-duction of WAP (*Wireless Application Protocol*, a protocol for adapting Internet accesses to the GSM network's limited bandwidth) technology and enhanced messaging services.

In this case as in that of multimedia services, having more available bandwidth opens new prospects which are far more interesting than the GSM system's offerings. At the moment, the Internet applications that are enjoying the greatest success are e-mail and access to infor-mation content such as sports news, financial bulletins, weather fore-casts and so forth.

The goal is to make all of this information usable from the mobile environment. With bandwidth limitations a thing of the past, it will no longer be necessary to specify new access protocols. However, as the characteristics of mobile terminals will continue to be profoundly different from those of a desktop PC – as regards computing power as well as the size and features of the display and keypad – it will still be necessary to use entirely different methods for viewing information content. The basic problem thus becomes that of redistributing information on the basis of access type, without having to rewrite the service each time.

At the time of access, the user will inform the service provider of his terminal's characteristics (display dimensions, computing capacity, graphic processing capacity, etc.), and the provider will transmit the required information, adapting representation to the indicated characteristics. Consequently, the same information will be presented in different ways, according to whether access is from a desktop computer or from a mobile terminal. To this end, new languages for specifying information and representing it graphically will be introduced which overcome the limitations of those used today, together with software platforms for integrated profile and service management, and visual programming environments for fast application development. The major software producers (IBM, Microsoft and Oracle) are already working on platforms of this kind for making Internet services available with GSM terminals.

As regards messaging services, which are one of today's most widely used applications – as witnessed by the success of e-mail or of the GSM network's short message service – the goal is to associate multimedia content with transmitted messages. For example, messages could be transmitted directly in the form of film clips.

7.3.4 Voice services

Services based on voice recognition and voice activation will also have an extremely important role. Today, in fact, one of the most formidable barriers to the success of value-added services on GSM consists in the telephone terminal's limited practicality for these purposes, and the complexity of its man-machine interface, or

MMI. This problem is not so much due to poor terminal design, as to the terminal's inherent nature.

Voice recognition technology could help overcome these limitations, at least for certain classes of service. For this purpose, the idea is now to develop a voice mark-up language called VoXML, which could provide a standard mechanism for controlling applications through voice commands.

Users would thus be able to access and control the various services (such as navigation-based services, for example) through voice instructions to the terminal or the network. For instance, the user could say 'Check my e-mail' to access incoming mail. Other similar services will become available. This would thus increase the usability of all of these applications, making them readily available even to the large group of customers who use the phone only for voice communication and are not well versed in the use of new technologies.

7.3.5 User identification and security

Smart card technology makes user identification possible. The basic concept is applied today in the GSM network, which uses an SIM module that can be placed in any terminal to identify the user to the network and permit access to services. A subscriber can use any GSM phone simply by inserting his or her SIM. This module thus makes it possible to introduce personalised features and security algorithms, thanks in no small measure to rapid advances in microchip construction and integration technologies that provide increased memory, interfacing and computing capacity. UMTS intends to make use of an identity module called a USIM, in which applications, certifications, digital signatures, encryption algorithms and any other type of data can be entered and stored. These cards could be contactless, which would make them easier to use and, above all, would extend their potential, as operations could be performed without necessarily having to insert or remove the card from the terminal.

Obviously, this will permit the large-scale introduction of commercial and financial transactions via UMTS for applications such as e-commerce, home banking, and so forth.

7.3.6 Location-based services

Services based on the concept of positioning are another sector of enormous interest, since mobility is the main characteristic of mobile radio systems and their principle source of added value. Even now, applications based on positioning systems are available on second-generation networks. In the current GSM network, a distinction can be made between two classes of application: those that use the mobile radio network only as a transmission medium, relying on specific systems such as GPS (*Global Positioning System*) for the actual positioning functions, and those that use the native techniques provided by GSM for locating the user (the cell identifier or, for higher accuracy, triangulation methods based on the different power levels received). The GSM system, however, is not organised to support positioning concepts on a complete, native basis. Consequently, despite concerted efforts and ongoing research, it has not yet been possible to develop applications of this second type which are in any way comparable in accuracy and reliability with the first type, which nevertheless have the disadvantage of requiring supplementary systems and equipment. UMTS should change this picture, and overcome this type of limitation. Some of the possible applications include:

- Work Force Management. A centralised system co-ordinates and monitors groups of geographically dispersed users (agents, repairmen, salesmen, etc.). At the moment, the GSM system is essentially limited to co-ordinating vehicles equipped with special on-board terminals that can interact with the GPS and GSM systems. With UMTS, the individual user will be able to utilise the terminal directly without additional equipment.

- Navigation, traffic control and theft-deterrent services. In addition to using UMTS to support on-board navigation, it would also be conceivable to monitor the movements of vehicles on which a UMTS positioning system has been installed. This would make it possible to transmit the vehicle's location to a service centre, reporting any problems such as accidents, calls for help, etc. By connecting this system to the vehicle's theft-deterrent system, moreover, it would be possible to track the position of stolen vehicles.

- Yellow pages services. A wide range of public services for every-day use could be provided. For example, users could request a list of pharmacies, restaurants and movie theatres in their area, together with additional information such as opening hours, prices, films being shown, etc. As another possibility, on-line tourist guides could be offered.

8

Equipment and Service Testing

Loris Bollea and *Valerio Bernasconi*

The testing on mobile radio systems has a number of objectives. From the manufacturers' point of view, prototype testing paves the way to equipment development, identifying any implementation difficulties beforehand. The standards-writing organisations can then work to solve the problems identified through these analyses. A mobile service operator, on the other hand, can begin to study the difficulties that will be encountered during network operation, well before the network is put into service. How well services will perform and what their impact will be on network operation can also be assessed. New services can then be tried out on sample groups of 'friendly' users to provide a preview of their customer appeal. Finally, testing enables both manufacturers and operators to design qualification procedures for commercial equipment.

For all of these reasons, there is enormous interest in carrying out tests with prototype UMTS systems. Of the many experimental activities which are already under way around the world, there can be no doubt that the most extensive is that conducted by NTT DoCoMo in Japan, which involves five complete prototypes. These prototypes were constructed by the major mobile radio equipment manufacturers and are capable of communicating with each other, i.e. network nodes made by two different producers can be connected.

In Italy, TIM and CSELT have co-operated in a test campaign based on equipment supplied by Ericsson. Other tests with Ericsson have been carried out in co-operation with major operators in Sweden, Great Britain, Germany, China and Japan. Nor have the other industries in the sector lagged behind. Those who have developed experimental prototypes, in fact, include Nokia, Motorola, Lucent, Siemens and the major Japanese manufactures.

Though they are experimental and provide simplified network functions, these systems allow to evaluate the services that can be provided to future customers from the qualitative and quantitative standpoints, and furnish pre-operational information for operators of a UMTS network. A further, and no less important result of studies on prototype systems is the support they give to the international organisations in standardising UMTS. The experimental systems installed by the principal mobile network operators around the world make it possible to assess the services that will be provided to customers starting from 2001 in Japan and 2002 in Europe. These services are not limited to the classic and popular voice telephony, but also include videoconferencing and multimedia services such as e-mail, image and data file transfer and, in general, all of the Internet applications that are now available from our personal computers in the office or home. From the operators' standpoint, the various experiments that have been carried out since 1998 were and are an effective means of evaluating network architectural solutions in the field, as well as the use of the radio resources that were adopted during UMTS system standardisation.

The tests that have been conducted in this initial period of UMTS system evaluation, moreover, are a good starting point for developing the acceptance and conformance tests that will be carried out on commercial equipment before it is put on the market. Like those now performed on GSM equipment, these tests will serve to determine whether units comply with international standards and specifications, and, essentially, must ensure that equipment constructed by different manufacturers is compatible. In general, the systems marketed by a producer of mobile terminals or network nodes are subjected to mandatory (regulatory) conformance tests (and to other supplementary tests where necessary) which are intended to ensure user safety, correct use of the radio spectrum and compatibility of the equipment under test with the rest of the network deployed in the field. In addition to mandatory tests, a number of quality aspects affecting equipment and terminals are evaluated using objective comparison criteria in order to select the products best suited for the single operator's network.

Two groups of tests are normally carried out on equipment. First, performance is measured in the laboratory, with a controlled radio environment. Other tests are then conducted in the field, i.e. on the road, with a real telecommunications system.

The tests carried out by accredited laboratories on second-generation mobile equipment (i.e. GSM) have been used as a basis for specifying a series of tests on the experimental system which are regarded as particularly significant in assessing performance in terms of the network and radio features and of the service profile provided to customers.

We will now have a detailed look at the Ericsson experimental system and the assessments that have been performed. In this historic moment, while the standards-writing groups are preparing the test specifications, the experimental system has been used as the test bench for initial validation of these standards.

8.1 The experimental system

The system is based on the W-CDMA (*Wideband direct sequence-CDMA*) access technique, with a chip rate of 4.096 Mchip/s in the 2 GHz band.

The experimental system used for tests has gone through several software and hardware releases. In general, there is a sharp distinction between software and hardware functions even in the GSM network equipment produced today by the various manufacturers. Network infrastructures must have a high degree of flexibility in order to facilitate the introduction of new end-user services and equipment management in a relatively short time, and without requiring their physical replacement. This goal can be achieved by updating the software used in the node processing and control blocks. The tendency is thus to choose highly flexible, easily re-configurable hardware with a large number of DSPs (*Digital Signal Processors*) and re-programmable logic units. Only when the new capabilities to be introduced on the network call for increased processing capacity the hardware is updated, often by replacing the circuit boards containing the microprocessors.

Close-up 8.1 – Flexible prototypes through re-configurable circuits

When a new telecommunications system is conceived, it is of fundamental importance to be able to set up testing and evaluation environments in the laboratory so that the system's operations can be verified, its performance assessed, and measurement methods developed. Nowadays, computer-aided design is widely used, making it possible to carry out simulations of a product's operation and future 'behavior' early in the design stage, even before a prototype is constructed.

In particular, thanks to the ever-increasing potential of baseband circuits, prototype designers make extensive use of readily re-configurable logic circuits such as FPGAs (*Field Programmable Gate Arrays*) and DSPs (*Digital Signal Processors*).

Because of their high flexibility, these systems are extremely useful for verifying the first drafts of new products, before all the hitches have been ironed out, and are equally useful for revising the functions of installed equipment directly in the field.

FPGAs are logic components that can be programmed directly by the user. There are a number of different FPGA architectures: each manufacturer has proposed different solutions. In general, an FPGA structure is made up of a large number of elementary cells connected to each other by means of programmable interconnection matrices. Each elementary cell consists of a combinatorial logic network followed by a set of flip-flops. Modern FPGAs reach densities up to 250 000 gates, though this number grows significantly every year.

DSPs are special microprocessors optimised for digital data

stream processing. They are programmed in Assembler language, or through high-level compiled languages. Prototypes making extensive use of re-configurable FPGAs and DSPs can be readily reprogrammed for debugging, or to facilitate the introduction of new capabilities. Prototypes can thus be adapted in order to evaluate different solutions at the time system specifications are being developed.

The services that the experimental system can provide are listed in Table 8.1. In addition to 8 kbit/s voice calls, they include high-speed UMTS services which are an innovation over GSM, with circuit oriented services up to 384 kbit/s and packet oriented services up to 470 kbit/s.

Table 8.1 Services provided by the experimental system

Call	Service
Mobile–Fixed	Voice, 8 kbit/s
Fixed–Mobile	Circuit oriented services at 64, 128, 384 kbit/s
	Packet oriented services up to 470 kbit/s
Mobile–Mobile	Voice, 8 kbit/s
	Packet oriented services up to 384 kbit/s

The system's capabilities, which were described at the theoretical level in the preceding chapters, are as follows:

- Three sectors for each BTS.

- Soft hand-over, i.e. management of the mobile terminal's movement between sectors belonging to two different BTSs.

- Softer hand-over, i.e. management of the mobile terminal's movement between sectors belonging to the same BTS.

- Hard hand-over, i.e. call management when different radio frequency carriers are used.

- Power control, whereby the lowest possible transmission power compatible with a predetermined quality of service can be used.

- Channel type switching, whereby radio frequency channel usage can be managed and optimised.

- Interconnection with the GSM network.

The experimental system, whose elements and the associated interconnections are represented in Figure 8.1, consists of the following network nodes:

- two Mobile Stations (MSs);

- two Base Transceiver Stations (BTSs);

- one Radio Network Controller (RNC);

- one Mobile Services Switching Center (MSC);

- one ES (Experimental System) management network.

As can be seen from Figure 8.1, the experimental system is a complete cellular telecommunications network in all respects, with all the various network controllers organised in a precise hierarchy. The structure can thus be effectively used to verify the behaviour of a new piece of equipment or a new service under all circumstances that could be encountered on the market.

The MSC's job is to interface with other telecommunications networks, starting with the Public Switched Telephone Network (PSTN), and then with other cellular networks such as the GSM network. The new feature for this system is the fact that it is directly connected with the Internet in order to provide multimedia services.

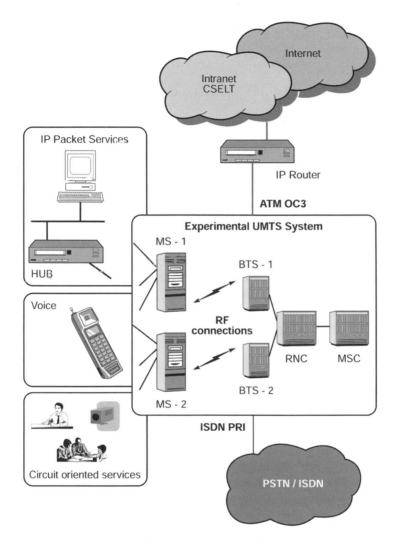

Figure 8.1 Experimental system connections for service provisioning.

As the UMTS system's switching node, the MSC manages and routes calls to several geographically dispersed RNCs.

The RNC carries out several of the functions that fall to the BSC in the GSM network, including BTS and radio resource management.

Finally, the BTS is the network node which interfaces directly with the user terminal. Its job, in fact, is to convert the information received from the switching nodes into a radio signal. The experimental system uses the following transmission frequencies:

- Uplink (transmission from the mobile terminal to the BTS): 1920–1940 MHz.

- Downlink (transmission from the BTS to the mobile terminal): 2110–2130 MHz.

The management network is used for the operation and maintenance procedures that are needed on a telecommunications network, as well as for monitoring and measuring network element operating parameters.

The two data networks on which the experimental system's operation is based can be seen in Figure 8.2: the management network

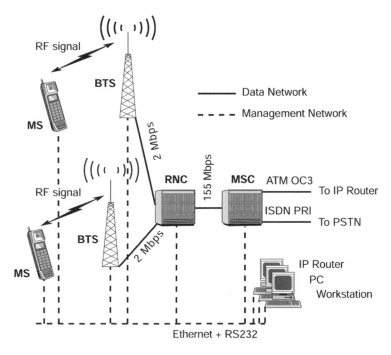

Figure 8.2 Network architecture for the experimental system.

described above, and the user data network. The latter uses 2 and 155 Mbit/s links on copper cable and optical fibre as the physical support for transmission.

In order to provide circuit oriented services, the experimental system is interconnected with the public switched telephone network, while IP-based services are accommodated through a connection to an external service provider.

8.2 Planned tests

The tests that have been carried out on the experimental system are only the first subset of those that will be performed on commercial equipment. In particular, each test is designed to simulate heavy use of a specific system function:

- Tests on physical layer aspects.

- Tests on system functions, processes and algorithms.

- Tests on network aspects.

- Tests on provided services.

All of these tests can be carried out in the laboratory, with a controlled radio environment, or in the field, i.e. under real cell coverage and operating conditions. The number of possible tests rises rapidly when certain variables are introduced in the system: the performance and operation of the algorithms on which the UMTS system is based, in fact, are heavily influenced by the speed of the transmitted information (the provided service's bit rate), the number of simultaneously active calls in a sector (the system load), signal propagation conditions in air (the propagation environment and mobile terminal speed) and by the reference parameters used for power control and hand-over algorithms. It will be up to the standardisation groups and operators to establish which combinations of tests are most effective in determining the performance and limits of the various systems.

Regarding issues affecting the physical layer, for example, tests have been conducted on the mobile terminal and BTS radio frequency signal emission masks. In addition, BER (*Bit Error Rate*) and FER (*Frame Erasure Rate*) measurements have been carried out under different system operating conditions and in relation to absolute values for received power and Eb/I0[1]. Network aspects are addressed by means of measurements and tests during hand-over procedures under different operating conditions, i.e. mobile terminal movement between two sectors belonging to the same BTS (softer hand-over), to two different BTSs (soft hand-over), or between different radio frequency carriers (hard hand-over). Interesting results regarding the feasibility and the objective and subjective quality of the provided service have been obtained by interconnecting the experimental system with other, existing, telecommunications networks such as the Inmarsat-B satellite network and the GSM network. These tests have been adapted from the corresponding ones performed on GSM network elements.

Tests on innovative customer services, on the other hand, were designed to assess the system's performance in terms of average IP packet transfer rate and transmission delay. Here again, results were obtained for different radio transmission parameters.

The third-generation systems will have to compete on the market with today's systems, which can offer innovative services based on the Internet protocol as well as high quality second-generation services. To evaluate the subjective and objective quality of voice services, which though hardly innovative are definitely fundamental, measurements and tests have thus been developed for estimating transmission delays of the voice signal, information losses when the received radio signal is low, echo and speaker sound quality.

---

(1) Eb/I0 is defined as the ratio of the energy of the BCCH (*Broadcast Control CHannel*) bit to the spectral density of the other signals received (e.g. the traffic channels and other interferers).

Close-up 8.2 – Test specifications for innovative equipment

At the time technical specifications were drawn up for the UMTS, the standards-writing organisations also laid the foundations for the specifications covering tests on the various types of equipment and their serviceability. For the physical layer in particular, operating parameters for the transmitter and receiver have been established. For the transmitter, for example, specifications have been prepared for output power dynamics (maximum value, power control algorithm speed), emission masks and modulation accuracy. For the receiver, on the other hand, sensitivity, interference immunity and the characteristics of the demodulation and synchronisation blocks have been defined. Alongside these specifications, UMTS network equipment measurement procedures for the radio parameters have been established that will be used both by accredited laboratories in order to test equipment, and by manufacturers of measurement instruments in order to develop innovative instruments, as will be illustrated in greater detail in the next Close-up.

Several 3GPP technical committees have drawn up specifications for the system and for equipment testing. For further information about the standards that are now being prepared, see the 3GPP Web site at www.3gpp.org.

8.3 Innovative services

If we look at the evolution of telecommunications and Information Technology, it is clear that several trends are already leading to convergence between telecommunications services and the Internet. This will certainly be possible (and indeed encouraged) with UMTS, as packet switched communication will be supported together with circuit switched services. At the current stage of development in this sector, we can even look forward to total connectivity in the future, when it should be possible to reach anyone, anywhere and transmit any kind of information in real time. In the next few years, it is likely that the explosion in the number of applications, multifunction terminals and technologies will bring data traffic volumes close to those for voice traffic. Considering these trends, the experimental system has already been used to verify the major services that could be provided to customers with the third-generation of mobile radio telecommunications services. These potential UTMS services include:

- Video-conferencing: these applications permit videophone-type communication between mobile terminals. The service has been tested using the MPEG-4 video coding technique (designed with wireless transmission systems in mind) and the IP protocols.

- Video-streaming: video recordings and images of various kinds can be sent and transmitted with this service. Customers will thus be able to receive real time television programs on their UMTS terminals for entertainment, cultural or educational purposes. Applications of this type have been tested on the experimental system with a 300 kbit/s average data transfer rate.

- Internet browsing: with this service, the UMTS network user can browse on the Web directly with his or her mobile terminal.

- Application sharing: applications running on the customer terminal can use processing resources resident on a remote server, e.g. one managed by the service provider.

As this brief overview has made clear, the innovative services provided by the UMTS system are closely linked to the IP world and are based on multimedia applications.

In the global mobility scenario now waiting in the wings, portable personal computers are not used only with standalone applications, but also and above all in a variety of mobility situations, maintaining a connection with their corporate network or service provider at all times. The multimedia applications that are now used to transmit information over the Internet or the fixed telephone network must be redesigned to make them compatible with mobility. Essentially, the restrictions hampering the use of today's multimedia applications in wireless applications include the limited processing capacity of a remote client, the limited bandwidth available for transmission, and, most importantly, the presence of errors introduced by transmission over the air interface.

In order to cope with the requirements that have recently arisen for mobile transmission, new audio and video compression algorithms have been developed which can operate with low information transmission rates (high compression for the starting video images) and low decoding algorithm complexity. Consequently, they can be used on standard portable personal computers. These algorithms also guarantee good recovery for errors introduced by mobility.

The experimental work now being carried out at CSELT is designed to determine the efficiency of the proposed solutions in a simulated or real mobility context, and to evaluate the performance levels that can be obtained from the transmission algorithms.

8.4 Laboratory testing

The measurement set-up depicted in Figure 8.3 is used to carry out the tests described in the foregoing paragraphs in the laboratory.

For the sake of simplicity, Figure 8.3 shows a single BTS and a single mobile terminal. The purpose of laboratory testing, where the BTS and the mobile terminal are connected by cable, is to assess a number of aspects involving not only the physical layer and the network, but also the potential services that can be provided by the UMTS system under controlled and repeatable signal propagation conditions.

For both the up-link (UL) and the down-link (DL), connection between transmitter and receiver is accomplished by means of cables and radio frequency components in order to emulate the effect of signal propagation on air. Interferers can be introduced to simulate system load by using arbitrary computer-controlled waveforms. For example, interferers can be generated by a computer through system simulation and then loaded in the waveform generator's memory in the form of files representing signal amplitude versus time.

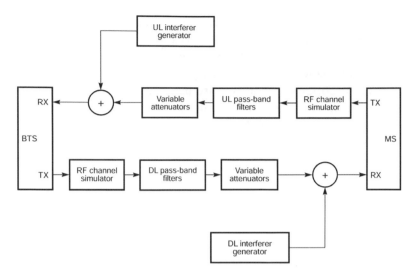

Figure 8.3   Connections for laboratory measurements.

By expanding the block diagram, summing the signals produced by several BTSs and controlling their amplitude, it is also possible to reproduce hand-over conditions between one sector and another with high precision. The mobile terminal's movement between adjacent sectors can be reproduced, monitoring the signal received by the mobile terminal arriving from a certain sector at all times.

An element of particular importance in the block diagram is the hardware simulator of the propagation channel. This extremely flexible and re-configurable instrument makes it possible to evaluate the behaviour of the system under test by simulating different propagation conditions for the signal on air. In the laboratory, the device under test can thus be fed with an input signal whose characteristics (such as amplitude and phase) are quite close to those that the device would receive in a real electromagnetic propagation environment. The characteristics of the propagation profiles are established by the standardisation groups. For UMTS, characteristics have been specified for the following propagation scenarios:

- Indoor

- Indoor to outdoor (large halls, courtyards, outdoor areas adjacent to buildings, pedestrian speeds).

- Vehicular (more extensive areas, traversed in a motor vehicle).

A parameter which can be readily modified through the channel simulator is the speed of the simulated mobile terminal.

Close-up 8.3 – Innovative measurement instruments

The introduction of wide-band systems employing CDMA access techniques has made it necessary to develop innovative instruments for testing equipment performance, as conventional spectrum analysers are not sufficient for the new measurement needs. A new family of vector signal analysers is thus coming to the market which combines the capabilities of traditional radio frequency instruments with computing resources comparable with those of modern workstations. These instruments first perform the appropriate frequency conversions, and then acquire a significant portion of the signal to be examined in their internal memory. This analysis takes place using algorithms such as the FFT (*Fast Fourier Transform*) for real time analysis of the signal spectrum, or by means of correlation with known code sequences for code domain power measurements (Figure 8.4).

As they are largely based on software, these instruments can adapt flexibly to analysing signals originating from different standards with both time division and code division techniques. These instruments measure channel power, modulation accuracy (ρ), code domain power and ACLR (*Adjacent Channel Leakage Power Ratio*). Alongside analysers, radio frequency signal generators are becoming increasingly configurable, making it possible to reproduce complex digitally modulated signals with enormous flexibility.

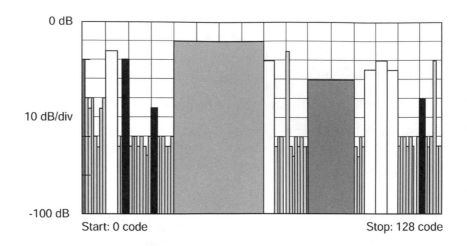

Figure 8.4   Code domain power measurements.

8.5 Field trials

Another innovative experiment carried out at the CSELT laboratories was the assessment of the pico cell coverage of areas adjacent
to the laboratory. It was thus possible to evaluate – with a relatively
high degree of quality and reliability – the type of UMTS services
that can potentially be provided to customers in an office environment, or, in other words, in a situation calling for high information
transfer rates but low mobile terminal speeds. The main achievement
in this stage of testing was the demonstration that IP-based high bit
rate services can be provided with a good level of connection reliability and with transmitting powers on the air which are in the same
order of magnitude as those used by current DECT telecommunications systems.

   The second stage of the field trial provided micro cell coverage in
Torino city centre. This made it possible to evaluate system performance and service reliability in an outdoor environment. Information
of major pre-operational importance was obtained, because the characteristics of the first urban areas in which the UMTS system will be

introduced are very similar to those investigated at this stage. In addition, this test also enabled researches to assess the connections needed for node deployment in the served area: the user data and system management networks illustrated in Figure 8.2 were not confined inside a laboratory as in the first stage of indoor testing, but were set up using branches of the public transport and distribution networks.

9

Research Topics

Enrico Buracchini

9.1 Introduction

This chapter discusses the topics that mobile communications research will be addressing in the near and immediate future. While mobile radio communication systems have seen a continuous rise in the number of users in recent years, further growth is hampered by the scarcity of available resources, and, in particular, by the limited spectrum of frequencies that can be used. Consequently, research centres on developing new access techniques which can make more efficient use of available frequency bands. In this connection, the *Space Division Multiple Access* (SDMA) technique makes it possible to increase the capacity of a cellular mobile radio system by taking advantage of spatial separation between users [4,23,24].

In an SDMA access system, the radio base station does not transmit the signal to the entire cell area, as in conventional access techniques, but concentrates power in the direction of the mobile unit for which the signal is directed, reducing it in the directions where other units are present. The same principle is applied in reception [4,23,24].

In addition, the exponential growth in mobile radio services has led over the years to the development of a plethora of analog standards, such as Europe's TACS and AMPS in the US, as well as digital standards such as GSM in Europe, D-AMPS (*Digital-AMPS*) and the IS-95 (*Interim Standard-95*) in the US, and PDC in Japan. As a result, a potential pragmatic solution is taking shape: a software implementation of the user terminal, capable of adapting dynamically to the radio environment surrounding it at any given time, and of being programmed for the particular standard used in the geographical area concerned.

9.2 The SDMA *access technique and smart antennas*

In conventional cellular systems, the radio base stations have no information regarding the mobile terminal's location, and are thus obliged to radiate the signal in all directions in order to cover the entire area of the cell. In addition to wasting power, this involves transmitting a signal which, in the directions where there are no terminals to be reached, will be seen as an interferer for the co-channel cells, i.e. the cells where the same group of radio channels is used. Similarly, in reception, the antenna picks up signals from all directions, noise and interference included. These considerations have led to the development of the SDMA (*Space Division Multiple Access*) technique, which is based on the idea of acquiring and using information regarding the mobile terminals' spatial position [4,23,24].

In particular, the technique involves dynamically adapting the base station antenna beam pattern for each different user. In this way, maximum transmitted power and reception gain is focused in the direction of the mobile unit considered; at the same time, these parameters will be nulled, or minimised, in the directions of the interfering units. This can be achieved by using an antenna array instead of a single omni-directional antenna (or a fixed directional antenna in the case of sectorisation). The signals from these antennas are converted into digital form and processed under the control of appropriate algorithms.

In practice, information about the characteristics of the interference environment is obtained from the signals received. This information is used on the reception side to coherently combine received signals, thus modifying reception gain in the various directions. In transmission, this information is used to determine the differences between the signals to be sent to the array elements, thus 'shaping' the radiation function. In the literature, antennas capable of the type of operation described above are called smart or adaptive antennas [1,17,26].

Typically, the beam pattern (Figure 9.1) features a main beam aimed in the direction of the source to be tracked, and a series of secondary beams (or lobes) whose characteristics depend on the type of algorithm used for control. In the directions of interfering mobile units, moreover, the beam pattern will show minima, or also said *nulls*. The resulting behaviour is called *null steering* [21].

The main advantage of the SDMA technique is system capacity increase, i.e. the number of users that can be managed simultaneously. Methods for achieving this increase will be discussed in the following paragraphs.

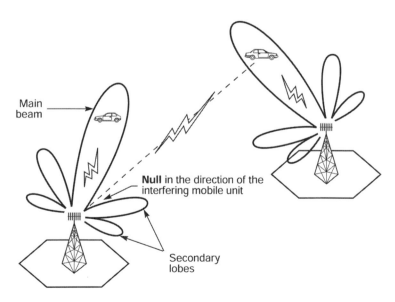

Figure 9.1 Example of null steering.

It is possible to exploit the higher receive gain offered by an antenna array with respect to an omni-directional case, to allow mobile units to transmit at reduced power, and therefore lower consumption. At equal power, antenna gain can be exploited to extend cell size. This is useful when it is necessary to cover vast surface areas, such as rural ones, characterised by a low mobile radio traffic density, with a limited number of base stations.

The use of an adaptive array can also be seen as an evolution of normal diversity techniques. In this case, rather than simply comparing the power level at the various antennas, it is possible to cancel multipath signals or, even, add the various contributions coherently, by adapting the antenna's radiation function appropriately.

The SDMA access technique can also be integrated with all of the other multiple access techniques (FDMA, TDMA, CDMA), and can thus be applied in any mobile radio system now in operation or introduced in the future. The necessary changes, in fact, are limited to the base station and do not affect the mobile units. It should be noted, however, that methods and advantages of introducing the SDMA technique will vary according to the characteristics of the system in question.

The SDMA technique can be introduced in gradual steps. It is, in fact, possible to initially modify only the base stations serving particularly critical traffic areas, without having to revise system planning as is necessary with traditional techniques such as cell splitting and sectorisation.

9.2.1 Applications of the SDMA technique

The SDMA technique can be introduced at various levels in a particular cellular system, taking advantage of the different performance features offered by an antenna array. The deployment strategy depends on the problems of the area in which the base stations to be modified operate.

Co-channel interference reduction

The most immediate effect of an adaptive antenna array is to reduce interference. This is possible because of the fact that power is no longer uniformly radiated in all directions by the base station, but can be focused only in the direction of the desired mobile unit. This signal causes interference with a mobile unit belonging to a co-channel cell only if the latter is located in the direction in which the beam is aimed. As there is a low probability that this will occur (and the likelihood depends in any case on beam characteristics) the average level of interference on the down-link is reduced.

Similarly, on the up-link, the base station receives only the signals coming from the direction of the desired mobile unit, and any other unit which happens to be is eventually using the same channel, will interfere only if it is located in this direction.

This technique is indicated with the acronym SFIR (*Spatial Filtering for Interference Reduction*), [4,23,24].

In Figure 9.2, for example, it can be seen that, on the down-link, only the interference due to base station BS1 reaches mobile unit M2 on the down-link, while, on the up-link, BS1 receives the interfering signal from M2. In the same example, if an omni-directional antenna were used, each mobile unit would always receive interference from two base stations, and each base station would receive that of two mobile units (assuming, obviously, that that same channel is in fact used in all three cells).

It should be specified that the objective of the adaptation algorithms used to cancel for interference cancellation is not to maximise gain in the direction of the mobile unit, but to maximise the carrier-to-interferer ratio C/I. This translates into *null steering* behaviour, meaning that reception and transmission radiation function nulling takes place in the directions where the interfering sources are located. In this case, the number of antennas in the adaptive array determines the system's degrees of freedom, and hence the number of interferers that can be eliminated.

The increase in the C/I ratio increase resulting from the use of an adaptive array makes it possible to reduce *cluster* dimensions, thus increasing the system's capacity.

At first glance, this application could appear to be an 'extreme' form of sectorisation, as using an array is equivalent to having a series of highly directive antennas. In reality, however, the differences between the two techniques are considerable. In the first place, the

C/I ratio increase achievable with the SFIR technique can be significantly higher than the one provided by sectorisation. Furthermore, the mobile unit is continually tracked by the antenna: in addition to reducing interference dynamically, this means that there is no need for hand-over procedures as the mobile unit moves from one antenna's range to another's, as in sectorisation. Consequently, all of the channels assigned to the base station can be made available over 360°, rather than in a single sector.

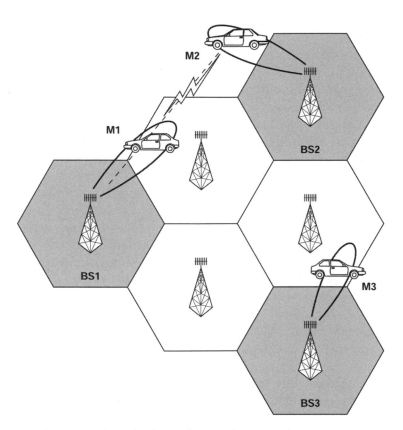

Figure 9.2   Spatial Filtering for Interference Reduction (SFIR).

Spatial orthogonality

Conventional access techniques are based on the possibility of making two signals *orthogonal* by allocating them in a different frequency band (FDMA), transmitting them at different times (TDMA), or assigning a different code to them (CDMA). Orthogonality between two signals ensures that the receiver can 'separate' them even if they are superimposed in the time and/or frequency domain upon reception.

The fact that two signals are transmitted to or from different directions can be turned to advantage in order to create a further degree of orthogonality [27]. It is thus possible to assign the same physical channel to several users (Figure 9.3), provided that there is sufficient

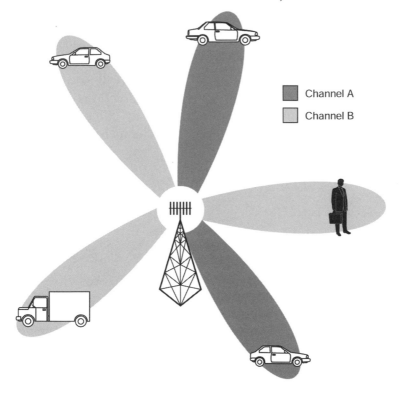

Figure 9.3 Spatial Division Multiple Access (SDMA); several users belonging to the same cell can use the same channel, as several beams can be created in different directions.

angular separation between them. This is the application to which the term Space Division Multiple Access is most correctly applied.

The obvious advantage of this technique is the increase in the number of available channels in a cell: each channel is no longer assigned to a single user, but is shared by a certain number of users. The parameter of greatest interest here is the *number of users that can be managed simultaneously in one channel*, defined as *Spatial Multiplexing Gain*, or SMG. In practice, the number of channels made available in cell N'_c is equal to SMG$^*N_c$, if N_c is the number of channels assigned to the base station [4].

Spatial multiplexing gain is closely linked to the number of antennas in the array. In fact, spatial orthogonality is exploited by using spatial filtering to eliminate *intracell co-channel interference*, or in other words, the interference due to the mobile units belonging to the same cell and using the same channel as of the considered mobile unit. The power level, with which the base station receives the signals transmitted by these mobile units, can be compared with that of the desired signal. It is therefore necessary that, as illustrated in Figure 9.4, the systembeam pattern presents, in the directions of all intra-cell interfering mobile units, very 'deep' nulls by comparison with the nulls that are sufficient to suppress co-channel interference from other cells.

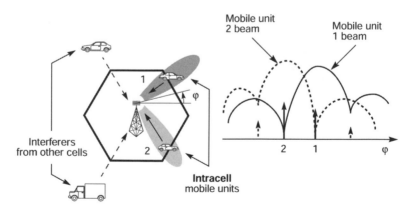

Figure 9.4 Beam pattern for two users employing the same channel: 'deep' nulls must be positioned in the direction of the intracell interferers.

To achieve this kind of selectivity in system response, the antenna array must have high directivity, and the number of antenna elements must thus be sufficiently large.

To effectively suppress the signal due to an interfering mobile unit, moreover, there must be sufficient angular separation between this unit and the desired unit. The minimum angle at which the two signals can be correctly separated also depends on the number of antennas. For this reason, when two co-channel intracell mobile units approach each other, one of them must be forced through appropriate signalling to 'abandon' the channel and use another on which there are no mobile units in the considered direction.

However, in addition to involving further implementation difficulties, increasing the number of antennas involves greater computing complexity of the adaptation algorithms, and generally translates into longer convergence times, thus compromising the correct track of fast moving mobile units.

Spatial Multiplexing Gain also depends on the users' spatial distribution. If users are uniformly distributed in the cell area, in fact, all degrees of spatial othogonality can be exploited. By contrast, if traffic is concentrated in a few directions, the number of users that can be managed for each channel is less than the maximum possible.

It should be noted that, strictly speaking, the acronym SDMA refers only to the last application described above, which is the only one in which 'spatial channels' are created and space division access is actually achieved. In current practice, however, the term is also used to designate all of the various applications of an adaptive array in a mobile radio system which take advantage of spatial separation between users.

An initial SDMA introduction, to increase spectral efficiency, can be limited to implementing SFIR, whereby cluster size can be reduced and more carriers provided to each base station. Real SDMA access, i.e. obtained by taking advantage of interference nulling to achieve spatial orthogonality, has the advantage of raising spectral efficiency directly by increasing the number of channels, and thus does not call for changes in the cluster structure. By contrast, full implementation of the SDMA technique involves several problems related to an higher complexity of necessary algorithms.

These considerations, however, apply only to systems with FDMA-

TDMA access. In a system with a CDMA code division multiple access, all users share the entire available frequency spectrum. Unlike systems with FDMA-TDMA access, system capacity is now limited only by the level of interference [9,13,18]. Any solution which reduces the interference level translates directly into increased capacity [3,16].

In a CDMA system, then, the SDMA technique can no longer be split between SFIR and full SDMA. Now, in fact, reducing interference with the SFIR technique directly increases the number of available channels, with no need to change the frequency reuse distance. In addition, it should be noted that in CDMA systems, the same frequency band is used in all of the system's cells, and the concept of frequency reuse thus no longer applies.

In a FDMA-TDMA system, there is a limited number of interfering mobile units for each user. These units' signals must be carefully eliminated in the SDMA technique by creating nulls in the corresponding direction of arrival. With the CDMA technique, as all mobile units share the same band, the number of potential interferers is very high, and will certainly exceed the number of antennas in the array, that is the number of degrees of freedom in the adaptive system. In addition, interferers can be considered to be uniformly distributed around the radio base station. This means that the beam pattern's adaptation system no longer exhibits null steering behaviour, and the pattern will show only the main beam aimed in the direction of the desired mobile unit [3,16]. In this way, the reduction in interference level will be only partial.

However, as the precision with which nulls are positioned in the direction of the interferers is no longer critical, the adaptation system requires a minor complexity.

Switched beams

When an adaptive antenna array is used, the mobile unit is 'tracked' constantly so that the main beam is aimed more or less exactly on it, and the nulls are positioned in the direction of the interferers. In another approach (Figure 9.5), a certain number of beams are generated and remain active at all times in reception. The system selects the beam from which the strongest signal is received, and uses it to trans-

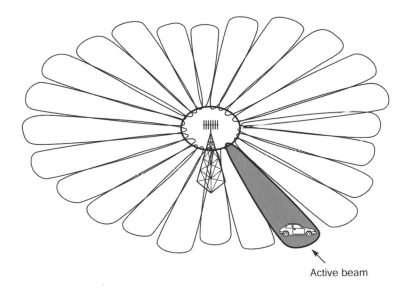

Active beam

Figure 9.5 Switched beams.

mit the signal to the mobile unit. As a result, this is referred to as the *switched beam* or *multibeam antennas* approach. Systems of this kind have been already proposed for GSM [8].

As regards increasing the C/I ratio, the system in this case can effectively be considered as an extreme form of sectorisation, as the cell area is divided into *microsectors*, each served by a single beam. There are a number of drawbacks to using this technique. Because the beams are fixed and have a predetermined radiation function, signal power varies as the mobile unit moves through the microsector. Furthermore, since only the received power level is considered, there is no distinction between the desired source and an interferer. If the latter is located in the direction of maximum beam gain, while the desired mobile unit is situated outside of it, the interference level may be increased.

Clearly, then, the advantages of multibeam antennas in terms of eliminating interference and reducing cluster size are less than those that can be obtained with the SFIR technique. In addition, it is not possible to implement the SDMA technique, i.e. to assign the same channel to several mobile units.

Nevertheless, the fact that this technique is less complex to implement means that it can be used in the mobile radio systems currently in service. The introduction of adaptive arrays, on the other hand, is expected to take place in the longer term.

9.3 Software radio

Since the early '80s, the exponential growth in mobile radio systems and services has led to the development of a large number of analog and digital standards around the world. With the new century, industrial competition between Asians, Europeans and Americans promises a very difficult path toward the definition of a single standard for future mobile systems, although market analyses underline the trading benefits of a common world-wide standard.

It is therefore in this field that the 'Software Radio' concept is emerging as a potential pragmatic solution: a SW implementation of the user terminal able to dynamically adapt to the radio environment in which it is, time by time, located. In fact, the term 'Software Radio' stands for 'radio functionalities defined by software', with this meaning the possibility to define by software, the typical functionality of a radio interface, usually implemented in transmitters and receivers by a dedicated hardware. The presence of the SW, defining the radio interface, entails the use of DSP processors, replacing dedicated hardware, in order to run the software in real time.

9.3.1 Software radio definition and its objectives

No rigorous definition of the concept of software radio has yet been advanced, though the need to clarify exactly what the term means has been voiced by many sources. A number of the definitions that are frequently found in the literature are given below [2,5–7,10–12,14, 15,19,20,22,25,28,29].

- Flexible transceiver architecture, controlled and programmed via software.

- Signal processing capable of replacing radio functions to the greatest possible extent.

- 'Air Interface Downloadability': dynamically re-configurable radio equipment by 'downloadable SW', at every level of the protocol stack.

- SW Implementation of 'multiple mode/standard' terminals.

- Transceivers in which the following can be established through software:

  – Radio channel frequency band and bandwidth;

  – Channel coding and modulation scheme;

  – Radio resource and mobility management protocols;

  – User applications.

Therefore, resuming, the following definition could be used:

'Software Radio is an emerging technology, thought to build flexible radio systems, 'multiservice', 'multi-standard', 'multi-band', re-configurable and re-programmable by software'.

The flexibility of a software radio system lies in its ability to operate in a multi-service environment without being bound by a particular standard, and at the same time being able to provide services in accordance with any of the systems that have already been standardised or will be standardised in the future, on any frequency band. The compatibility of a Swradio system with any defined radio mobile is guaranteed by its re-configurability, i.e. by DSP (Digital Signal Processing) engines re-programmability, that, in real time, implement radio interface and upper layers protocols. It is important to bear in mind that the term DSP (*Digital Signal Processing*) applies to the whole concept of digital signal processing, and thus covers FPGAs (*Field Programmable Gate Arrays*) and general purpose processors such as INMOS transputer and INTEL PENTIUM/MMX, as well as DSP chips *per se* [15].

From today's perspective, we are still nowhere near being able to construct a software radio system because of the many problems, most of which are technological in nature, which, at the moment, seem neither easy to solve nor likely to be solved soon. Nevertheless, software radio as defined above is a major research goal which we hope to succeed in reaching by 2005–2010.

Two main milestones must be passed first on the road to constructing a software radio system [14,15]. They consist of:

1   Moving the border between the analog and digital worlds in transmitters and receivers closer to radio frequency (RF) by adopting wide-band A/D (Analog to Digital) and D/A (Digital to Analog) converters which are as close as possible to the antenna.

2   Replacing ASICs – *Application Specific Integrated Circuits* – technology (dedicated hardware) with DSP (*Digital Signal Processing*) technology for base band signal processing so that radio functions can be defined by software to the greatest possible extent.

In reality, the first of these objectives is not exclusive to software radio, as witnessed by the years of efforts that have been put into implementing the so-called *Wideband Transceivers*, whose main goal is to expand the digital world as far as the IF (*Intermediate Frequency*), leaving only the RF part analog. Ultimately, the aim is construct all-digital transceivers.

Software radio attempts to incorporate the achievements made in the field of wide-band transceivers, and to go beyond them by introducing the possibility of reprogramming the entire system.

Replacing ASICs technology with DSP technology opens up two possible horizons:

1   Software implementation of base band functions such as coding, modulation, equalisation, pulse shaping, and so forth.

2 System re-programmability in order to guarantee multistandard operation.

Currently, ASICs is the predominant technology used in transceiver and receiver construction, and the circuital solutions adopted by the various manufacturers are highly proprietary. In any case, the use of DSPs is well-established, as UMTS radio base station prototypes have already been produced which rely heavily on DSPs for base band processing. Speaking of software radio systems in these cases is premature, as not all of the base band functions are implemented on DSP. In addition, the software is limited and preloaded, so that the entire system is bound to a specific radio interface with no provision whatsoever for reconfiguration.

Close-up 9.1 – Software radio transceiver

The transmitters and receivers, currently used in cellular mobile radio systems, are based on the traditional superheterodyne scheme (Figure 9.6), where the RF and IF stages are entirely analog and the digital component is present only in the base band (BB) stage, which is normally constructed in ASICs technology, i.e. with dedicated circuits. The only exceptions are the first-generation analog systems (TACS, AMPS), whose transceiver is all-analog.

By contrast, in the 'ideal' layout of a software radio transceiver, the analog stage is reduced to the bare essentials (Figure 9.7): the only analog components are the antenna, the bandpass filter and the low noise figure amplifier. Analog-to-digital conversion is carried out immediately at radio frequency so that all-digital signal processing can be performed on a fully re-configurable card.

The Software Radio receiver illustrated in Figure 9.7 has been defined 'ideal', since there are several matters that make it, at the moment, so far to be realised. First, it would not be conceivable to use a single RF stage for a multi-band system, as low noise figure amplifiers and antennas cannot be constructed for an operating bandwidth ranging from hundreds of MHz to units or tens of GHz. Today, the only way to guarantee multi-band operation capability is to construct several RF stages, depending on the number of bands used in the software radio system.

The most promising solution, at least at the moment, is known as 'Digital Radio transceiver', whose receiver section is shown in Figure 9.8. His structure is very similar to the 'Wideband transceiver' one, with RF stage completely analog and the digital extending towards IF.

The ADC (*Analog to Digital Conversion*) unit samples the entire spectrum allocated to the system, while the *Programmable Down Converter* carries out the following operations:

1 *Down Conversion*: digital conversion from intermediate frequency (IF) to base band (BB) by means of a table containing samples of a sinusoidal carrier. The table replaces the local oscillator used for analog down converters.

2 *Channelisation*: selection of the carrier to be processed by means of a digital filtering operation. In analog receivers, this operation is carried out using an analog filter with stringent specifications prior to BB conversion.

3 *Sample Rate Adaptation*: sub-sampling of the channelisation filter output signals in order to adapt the sample rate to the bandwidth of the selected signal, which is a narrow-band signal by comparison with the full spectrum signal at the input of the ADC converter.

The digital signal output from the IF stage is then subjected to base band processing.

Constructing a Digital Radio transceiver, however, involves considerable difficulties in both the IF and BB stages. For the IF stage, as, of course, for the RF one, the problems are essentially technological in nature, and are associated with the performance of the A/D and D/A converters and the resulting trade-off between sample rate and resolution: the higher the sample rate, the lower the number of bits with which samples can be represented will be. Present-day technology, for example, makes it possible to reach a sample rate of 1 GS/s with a resolution of 6–8 bits, dropping to 100 MS/s with 10 bits, and 150 kS/s with 16 bits [12,28]. The number of bits may not be sufficient if we bear in mind that the signals to be sampled may have very wide dynamic ranges. A GSM signal, for example, may have a dynamic range of over 100 dB. Other problems yet to be resolved include those associated with converter bandwidth limitations, frequency jitter, and the presence of inter-modulation products affecting the sampled signal. A rough estimate for obtaining a valid representation of the GSM and UMTS wave-forms and the associated dynamic ranges, confirmed in the litera-ture, could call for a resolution of at least 17–20 bits, with an analog sampling bandwidth of 250–300 MHz [12,28].

Technological problems also exist in the base band, where they are associated with DSP processing power and power consumption. Sufficient processing power must be available to allow real-time execution of SW implemented radio interfaces. This may call for the use of several processors in parallel, depending on the complex-ity of the radio interface to be implemented. As the software radio system must be able to adapt to different standards, it is necessary to size processing power for worst-case conditions. Technical solu-tions adopting multi-user detection (*for TD-CDMA based systems*) or beam-forming algorithms (systems with adaptive antennas), lead to an exponential growth in the required processing power, which, for the UMTS system, can be estimated in units-tens of GIPS (Giga Instructions Per Second) [25]. In addition, processing power is heavily influenced by the type of capabilities required of the DSP. General Purpose (GP) DSPs, in fact, are suitable for source and channel coding, encryption and modulation. At the moment, Special Purpose (SP) DSPs are necessary for carrying out complex real time operations, such as frequency conversion, filtering, spreading and despreading, which call for a processing power of at least 1200–1500 MIPS (Mega Instructions Per Second), which is currently available only in SP processors [2,11,12]. Nevertheless, GP processors increase their processing power every year, and some units are already capable of reaching a thousand MIPS. For exam-ple, TI (Texas Instruments) is developing the TMS320C6X and

TMS320C7X fixed and floating point families with 32-bit MAC (Multiplier/Accumulator). The 6X family executes eight instructions in parallel with a 5 ns cycle and can reach 1600 MIPS. The 7X floating point family is designed to reach 1 GFLOPS (Giga FLoating point Operations Per Second). Analog Devices, on the other hand, is developing a 32-bit floating point Sharc DSP capable of reaching 1500 Mflops, and a 16-bit fixed point version capable of reaching 4 GIPS [2,11,12].

Another parameter to be optimised in the architectural choices for these DSPs is the number of I/O operations and external memory access operations [2,11]. As for a PC, in fact, accessing the external bus reduces the effective number of operations per second by a few orders of magnitude, and becomes the real bottleneck for the entire system, neutralising all of the efforts made to achieve high DSP processing powers. In any case, there are precise constraints on the use of DSPs in base band processing. These constraints, which are particularly stringent if considered from the standpoint of the mobile terminal, include:

- limited circuital complexity;

- low cost;

- low power consumption;

- compact transceiver dimensions.

Besides, in BB stage, design issues have to be solved, related to the optimal choice of HW and SW architectures to be adopted (HW/SW partitioning and co-design).

Finally, a possible evolutionary pathway for a Swradio terminal could consist of the following three steps [10,20]:

- STEP 1: the terminal is capable of processing the source and channel coding via software (with DSPs or FPGAs), while the base band MODEM parts employ digital ASICs technology and the RF/IF parts are implemented with conventional analog circuits.

- STEP 2: in addition to the source and channel coding, the terminal is capable of processing the base band MODEM parts via software (with DSP or FPGAs).

- STEP 3: in addition to the source and channel coding and the base band MODEM parts, the terminal is capable of processing the signal at intermediate frequency (or directly at RF) obtained from a broadband RF stage, followed by a high-resolution, high-speed conversion stage.

The technical issues for complete implementation of a Swradio terminal are summarised in Figure 9.9.

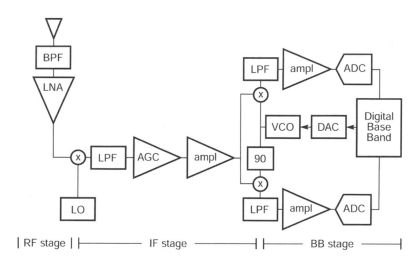

Figure 9.6 Conventional analog receiver scheme.

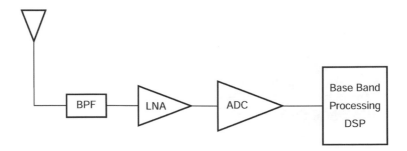

Figure 9.7   Ideal software radio receiver scheme.

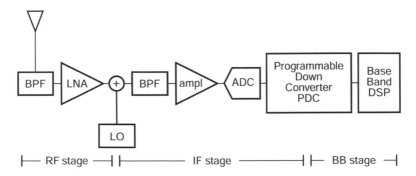

Figure 9.8   Digital Radio receiver scheme.

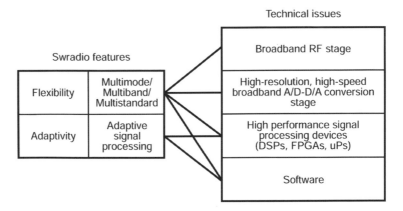

Figure 9.9   Technical issues for Swradio terminal implementation.

9.3.2 Possible solutions for software radio implementation

As mentioned above, a software radio system must be able to adapt to a wide variety of systems that have already been standardised or will be standardised in the future, by means of a common hardware platform, adaptable to any radio interface by simply changing the software running on it (Figure 9.10).

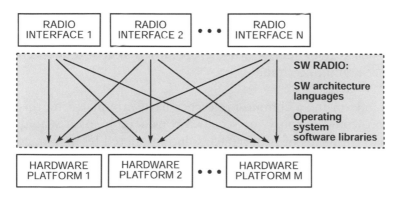

Figure 9.10   Hardware and software synergies in software radio systems.

The hardware design is carried out in parallel with the software one. Firstly, this requires the choice of a proper software architecture (likely object oriented), with its development environment and programming language. The final goal is the design of software libraries that implement the considered radio interfaces.

It is likely that the various manufacturers will develop proprietary products, or, in other words, different hardware platforms which are not compatible with each other. Consequently, software will be the substrate which guarantees that the radio interface is independent of the hardware platform. It is possible to identify three solutions to develop a software radio system [6,19,25,29].

1   Developing specific proprietary software for each hardware platform: each manufacturer proposes its own hardware/software solution independently of the others, with the advantage of

being able to put a product differing from the competitors on the market. Which products are successful and which are not will then be decided by market forces. The software produced through this approach is optimised for the hardware on which it runs. The main drawback of this solution lies in the burdens that must be shouldered by the operators, who will have to deal with non-standard and hence highly proprietary products, and will also have to manage the software to be loaded on user terminals. Note that the user is perfectly free to buy any brand of terminal, and the operator must then load that terminal with software which is both specific for the operator's system and specific for that brand of terminal.

2   Standardising a common reference hardware platform: this solution would simplify the scenario considerably, as it would eliminate all forms of proprietary solution offered by manufacturers. Not only would there be a single type of hardware, but only one type of software would have to be developed ("one code fits all"). In this case, the main disadvantage is for the manufacturers, who would not be able to differentiate their products from those of the competition.

3   Using resident compilers and/or a standardised real time operating system: the compiler produces an executable code which is specific for the hardware on which it must run, starting from a source code that may be common to each platform. This software approach is thus referred to as 'write once, run anywhere'. This is the solution that has attracted the greatest consensus, as it incorporates all of the advantages and none of the drawbacks of the two preceding approaches.

One of the candidate languages for developing software in a software radio context could be JAVA, as its development philosophy is compatible with the objectives that software radio sets out to attain, though modifications would have to be made for real time operation requirements. Confidence in JAVA's success as a development environment for many user applications is running high in a number of quarters: Texas Instruments, for example, has already produced

DSP processors capable of interpreting JAVA byte code instructions [29].

More than a language, JAVA is a development environment based on an object oriented programming language whose syntax is very similar to C++. The basis principle of JAVA is that the applications must reside on remote systems, whereas, in local systems the resident software is reduced in order to limit the memory size. When needed, the application software can be downloaded from the network (e.g. Internet) and stored until it serves. The basic requirement for this philosophy's success is that it must be possible for the software resident on remote systems to be downloaded and run on any local system, independently of the type of processor or operating system used in the local system. This task is performed by a restricted core of instructions ('JAVA KERNEL') resident in the local system to interpret the application downloaded from the network. To give a concrete example, this core is resident in browsers such as Netscape or Explorer. In practice, JAVA's philosophy is write once, run anywhere.

JAVA applications are first precompiled in order to produce the so-called JAVA byte code, which is not a machine code, but a higher level code which is extremely compact and thus easily stored and transported on the network (Figure 9.11).

The JAVA byte code can be downloaded from the network and subjected to a final compilation stage by means of a resident compiler (in reality, JAVA uses an interpreter) to generate the executable code. The resident compiler constitutes what is called the JAVA virtual machine, which guarantees that the software is independent of the hardware platform on which it must run.

Bearing these concepts in mind, the software radio system can be seen as a layered structure as shown in Figure 9.12. The system is entirely defined by software, from the physical layer functions up to the application layer. The presence of a resident compiler ensures that software is independent of hardware.

A fundamental part of this layering process is standardising the APIs (*Application Programming Interfaces*), i.e. the interfaces between the application and the protocol stack used (high level API in Figure 9.12) and between the protocols and the HW/circuital component (low level API in Figure 9.12), called the *Virtual*

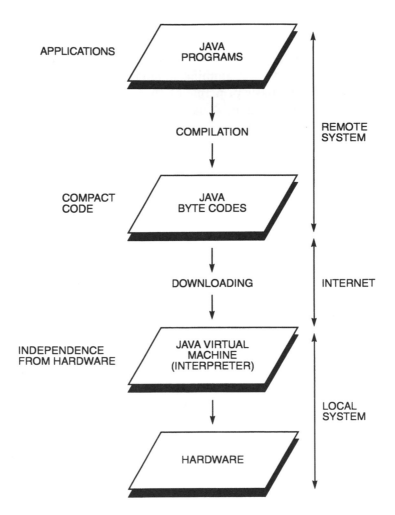

Figure 9.11   Execution of a JAVA application.

Machine or *Virtual Radio Platform*. As we have said, this interface would make it possible to support re-programmability for the lower layers of the protocol stack by providing a level of abstraction which conceals the actual HW implementation for the radio equipment. Standardising this virtual platform would also make it possible to download SW in terminals implemented using different

PROGRAMMABILITY

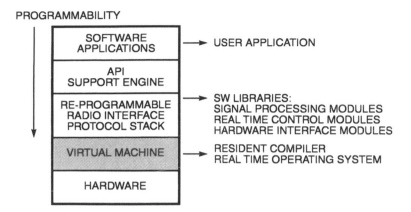

Figure 9.12 Relationship between hardware and software in a software radio

HW solutions and SW microcodes. This would enable manufacturers to continue to compete on the market by developing advanced proprietary solutions and standardising only certain minimum HW requirements like, e.g. transmitted and received powers MIPS and memory size of DSPs, converter resolution bits, display characteristics, and so forth [19,25].

It is important to notice that API is an abstract definition: API is not a particular code, program, application, but a description of the logical relationships among related SW and/or hardware modules, such as the bi-directional flow of data and control information. API describes the relationships of modules, not their implementation: the interfaces should be independent from the particular implementation [7].

The software defining the radio interface, together with the next-higher layers (layers 2 and 3), is collected in libraries containing the various modules. From the functional standpoint, these modules can be grouped into three categories:

1 *Signal processing modules*: these are the modules that implement basic functions (coding, modulation, etc.).

2 *Real time control modules*: these are the modules that superintend

the general operation of the entire process, calling and scheduling the signal processing and hardware interface modules.

3 *Hardware interface modules*: these are the modules that manage data input and output to the IF stage.

9.3.3 Software download

Implementing a fully re-configurable radio system entails specifying a method for loading software. Loading the radio base station is not a problem, as it can be carried out at the time the network is set up, in the same way that normal exchange software is installed. The software can be replaced whenever a new version is released, but this does not occur frequently.

The critical point consists of loading user terminal software, given that the terminal is the entity that must adapt to the system providing radio coverage. As a result, software loading may be required fairly frequently. Secondly, loading must be accomplished in as little time as possible, and involve the least possible amount of trouble for the user.

The two major methods for loading software are as follows [5,7,19,22,25]:

- Smart Card Loading (e.g. SIM);

- Over the Air Downloading.

Smart card loading

The user asks the operator, with whom he or she has subscribed to a contract, for a smart card containing the software, available from one of the operator's sales outlets or customer service centres. Once the smart card is inserted, the terminal loads the software defining the system used by the operator. Problems may arise if the user needs to roam on another network in the same country, or abroad, belonging to an operator who uses another system. In this case, the user must have several smart cards and know which one to use at any given time. The most likely alternative, however, is to enable the user to reprogram the smart card upon arriving in the host region or country by means of appropriate machines located, for example, at airports, stations, hotels or operator dealerships.

The advantages of this solution are as follows:

- Software integrity is assured, unless the smart card is damaged.

- Software can be loaded quickly, in times equivalent to those required for an ordinary PC.

- No impact on the network.

The main disadvantage is that the loading procedure is not transparent to the user. In other words, the mobile terminal cannot be reprogrammed unless the user does it himself.

In order to implement this solution, two requirements must be satisfied:

- Smart cards must be developed which have sufficient capacity to contain the software (technical requirement).

- There must be an efficient network for distributing smart cards, or machines for programming them (organisational requirement).

Over the air downloading

In some ways, this solution is the exact opposite of the previous one. Software is downloaded from the network via a dedicated service channel. The entire procedure is managed and controlled by the terminal and by the radio base station.

The main advantage of this solution consists in the fact that the downloading procedure is completely transparent to the user. The user may not even be aware of what is happening, as it is the terminals' job to identify its presence in a mobile radio system area and manage the software downloading procedure. At times, however, requests to load software or a portion thereof may be initiated by the user: for example, the user may ask to install a new video coder or additional applications. In such cases, though the user knows what is happening, all he or she need do is make a simple downloading request, and the mobile terminal and base station manage the rest of the process between them.

On the debit side, this solution has the following disadvantages:

- It has a heavy impact on the network, in the sense that it is necessary to set aside a dedicated channel for software download and to define a downloading procedure.

- Software integrity may be compromised despite the use of channel coding and/or packet repetition techniques, there can still be a problem with software integrity due to errors introduced by the radio channel.

- The downloading procedure can be slow: downloading time will depend on the size of the software and on the transfer rate permitted by the channel, i.e. on its bandwidth. There is a risk that the duration of the entire procedure will be far from negligible.

- Security of the authentication procedure for the terminal and for the entity which must download software.

For this solution to be successful, it will be essential to draw up a world-wide standard for a bi-directional communication channel for downloading software, which has sufficiently high capacity to keep downloading time within acceptable limits.

9.3.4 Benefits of software radio

Software radio can bring benefits for all of the players on the telecommunications scene, be they manufacturers, operators or users [5,22,25].

For manufacturers, it will be possible to concentrate research, design and production efforts on a reduced hardware platform set, which can be used in any kind of system and sold on any market, rather than only on the national or regional markets. With the enormous economies of scale this provides, it cannot fail to bring production costs down. A further and far from insignificant advantage consists in the ability to improve the software at subsequent stages, and to improve any errors or shortcomings which come to light later.

Operators can count on fast service rollout, and will be able to differentiate their services from those of other competing operators and provide new ones without having to change the user terminal each time – an advantage which will be shared by the user. In addition, it will be possible to implement several standards on the same radio base station.

Users will be able to roam with different systems, with the advantage of being able to enjoy world-wide mobility and minimise the risk of finding themselves outside of a coverage area: all that is necessary to provide the service is that there be a system of some kind which guarantees coverage. Nor should we forget that the user will be able to use terminals which can be configured on the basis of the applications and modes that he or she prefers.

In addition, software radio extends hardware lifetime for both the base station and the user terminal, pushing obsolescence farther into the future. The system's re-programmability allows hardware to survive for a longer period, and enables it to be reused when new-generation systems make their appearance. This does not mean that a user terminal will last forever: as the history of personal computers has shown, ever-higher computing powers are needed to run ever more powerful programs, and the computers of the previous generation soon fall victims to obsolescence. Though SW Radio terminals will have a longer lifetime than their conventional counterparts, they can expect to be affected by the same cycle.

References

[1] Anderson S., et al., 'An Adaptive Array for Mobile Communication Systems', *IEEE Trans. Veh. Technol.*, Vol. 40, No. 1, February 1991.

[2] Baines R., 'The DSP Bottleneck', *IEEE Comms. Mag.*, May 1995.

[3] Buracchini E., et al., 'Performance analysis of a mobile system based on combined SDMA/CDMA access technique', IEEE ISSTA '96, Mainz, 22–25 September '96.

[4] Buracchini E., et al., 'Accesso Multiplo a Divisione di Spazio in sistemi radio-mobili', *CSELT Technical Reports*, Vol. XIV, No. 1, February '96.

[5] Buracchini E., 'SORT & Swradio Concept', ACTS MOBILE SUMMIT 1999, 8–11 June 1999, Sorrento, Italy.

[6] Carpenter P., et al., 'Implementing Terminal Configurability in the Network', Software Radio Workshop, Brussels, May 1997.

[7] Cummings M., et al., 'MMITS Standard Focuses on APIs and Download Features', Wireless Systems Design, April 1998.

[8] 'DCS1800 Now Gets Active Across Europe', Microwave Engineering Europe, January '94.

[9] Gilhousen K. S., et al., 'On the Capacity of a Cellular CDMA System', *IEEE Trans. Veh. Technol.*, Vol. 40, No. 2, May 1991.

[10] Huomo H., 'Software Radio, a Manufacturer's Perspective', 1st International Software Radio Workshop, Rhodes, June 1998.

[11] Kostic Z., 'DSPs in Cellular Radio Communications', *IEEE Comms. Mag.*, December 1997.

[12] Kraemer B., et al., 'Advances in Semiconductor Technology Enabling Software Radio', Software Radio Workshop, Brussels, May 1997.

[13] Lee W. C. Y., 'Overview of Cellular CDMA', *IEEE Trans. Veh. Technol.*, Vol. 40, No. 2, May 1991.

[14] Mitola J., 'The Software Radio Architecture', *IEEE Comms. Mag.*, May 1995.

[15] Mitola J., 'Software Radio Technology Challenges and Opportunities', Software Radio Workshop, Brussels, May 1997.

[16] Naguib A. F., Paulraj A., Kailath T., 'Capacity Improvement with Base-Station Antenna Arrays in Cellular CDMA', *IEEE Trans Veh. Technol.*, Vol. 43, No. 3, August 1994.

[17] Nørklit O., Andersen J. B., 'Mobile Radio Environments and Adaptive Arrays', PIMRC '94.

[18] Pickholtz R. L., et al., 'Spread Spectrum for Mobile Communications', *IEEE Trans. Veh. Technol.*, Vol. 40, No. 2, May 1991.

[19] Robinson B., 'Software Radio: the Standards Perspective', Software Radio Workshop, Brussels, May 1997.

[20] Shinagawa Y., 'Software Radio Technologies', 1st International Software Radio Workshop, Rhodes, June 1998.

[21] Swales C., et al., 'The Performance Enhancement of Multibeam Adaptive Base-Station Antennas for Cellular Land Mobile Radio Systems', *IEEE Trans. Veh. Technol.*, Vol. 39, No. 1, February 1990.

[22] Swinburne M., 'Software Radio: an Operator's Perspective & Market Trends', Software Radio Workshop, Brussels, May 1997.

[23] Tangemann M., et al., 'Introducing Adaptive Array Antenna Concepts in Mobile Communication Systems', Proceedings of the RACE Mobile Communications Workshop, Amsterdam, May 1994.

[24] Tangemann M., Rheinschmitt R., 'Comparison of Upgrade Techniques for Mobile Communication Systems', Proceedings of the SUPERCOMM/ICC '94, New Orleans, Los Angeles, Vol. 1, May 1994.

[25] Tuttlebee W., 'The Impact of Software Radio', Software Radio Workshop, Brussels, May 1997.

[26] Van veen B. D., Buckley K. M., 'Beamforming: A Versatile Approach to Spatial Filtering', *IEEE ASSP Mag.*, April 1988.

[27] Weis B. X., 'A Novel Algorithm for Flexible Beam forming for Adaptive Space Division Multiple Access Systems', PIMRC '94.

[28] Wepman J., 'A/D Converters and Their Applications in Radio Receivers', *IEEE Comms. Mag.*, May 1995.

[29] www.ti.com/wireless, 'TI & JAVA: Bringing the Revolution in Network Intelligence to Wireless Systems'.

Acronyms and Abbreviations

3GPP	3rd Generation Partnership Project
A/D	Analog to Digital
AAL 2	ATM Adaptation Layer 2
AAL	ATM Adaptation Layer
ACeS	Asia Cellular Satellite
ACLR	Adjacent Channel Leakage Power Ratio
ADC	Analog to Digital Conversion
AICH	Acquisition Indication CHannel
ALCAP	Access Link Control Application Protocol
AMPS	Advanced Mobile Phone Service
API	Application Programming Interfaces
ARIB	Association of Radio Industries and Business
ARQ	Automatic Repeat reQuest
AS	Active Set
ASICs	Application Specific Integrated Circuits
ASU	Active Set Update
ATM	Asynchronous Transfer Mode
BB	Base Band
BCCH	Broadcast Control CHannel
BCH	Broadcast CHannel
BER	Bit Error Ratio
BG	Border Gateway
BGP	Border Gateway Protocol
BPSK	Binary Phase Shift Keying
BR	Border Router
BS	Base Station

BSC	Base Station Controller
BSS	Base Station Subsystem
BTS	Base Transceiver Station
CCCH	Common Control CHannel
CCPCH	Common Control Physical Channel
CDMA	Code Division Multiple Access
CID	Channel Identifier
CN	Core Network
CPCH	Common Packet CHannel
CPICH	Common Pilot CHannel
CRC	Cyclic Redundancy Code
CS	Circuit Service
CSCF	Call State Control Function
CWTS	China Wireless Telecommunication Standard
D-AMPS	Digital-AMPS
D/A	Digital to Analog
DCCH	Dedicated Control CHannel
DCH	Dedicated CHannel
DHCP	Dynamic Host Configuration Protocol
DECT	Digital Enhanced Cordless Telecommunications
Diff-Serv, DS	Differentiated Services
DPCCH	Dedicated Physical Control CHannel
DPDCH	Dedicated Physical Data Channel
DSCH	Downlink Shared CHannel
DSP	Digital Signal Processor
DTCH	Dedicated Traffic CHannel
EDGE	Enhanced Data rate for the GSM Evolution
ERAN	EDGE Radio Access Network
ES	Experimental System
ETSI	European Telecommunications Standard Institute
FA	Foreign Agent
FACH	Forward Access CHannel
FBI	FeedBack Information
FCC	Federal Communications Commission
FDD	Frequency Division Duplexing
FDMA	Frequency Division Multiple Access
FER	Frame Erasure Ratio
FFT	Fast Fourier Transform

FPGA	Field Programmable Gate Arrays
FPLMTS	Future Public Land Mobile Telecommunication Systems
GFLOPS	Giga FLoating point Operations Per Second
GGSN	Gateway GPRS Support Node
GIPS	Giga Instructions Per Second
GMPCS	Global Mobile Personal Communications via Satellite
GP	General Purpose
GPRS	General Packet Radio Service
GPS	Global Positioning System
GSM	Global System for Mobile Communications
GSN	GPRS Support Node
GTP	GPRS Tunneling Protocol
HAAP	High Altitude Aeronautical Platforms
HALO	High Altitude Long Operation
HLR	Home Location Register
HO	HandOver
HPLMN	Home Public Land Mobile Network
HSCSD	High Speed Circuit Switched Data
HW	Hardware
IETF	Internet Engineering Task Force
IF	Intermediate Frequency
IMSI	International Mobile Subscriber Identity
IMT-2000	International Mobile Telecommunications-2000
INMARSAT	INternational MARitime SATellite Organization
Int-Serv, IS	Integrated Services
IP	Internet Protocol
ISDN	Integrated Service Digital Network
ISL	Inter-Satellite Link
ISP	Internet Service Provider
IS-95	Interim Standard-95
ITU	International Telecommunication Union
LA	Location Area
LAN	Local Area Network
LEO	Low Earth Orbit
LES	Land Earth Station
MAC	Media Access Control
MAC	Multiplier/Accumulator

MAP	Mobile Application Part
MEO	Medium Earth Orbit
MES	Mobile Earth Station
MexE	Mobile Execution Environment
MGCF	Media Gateway Control Function
MGW	Media Gateway Function
MIP	Mobile IP
MIPS	Mega Instructions Per Second
MRF	Multimedia Resource Function
MS	Mobile Station
MSC	Mobile Switching Centre
MSISDN	Mobile ISDN number
MSS	Mobile Satellite System
NBAP	Node B Application Protocol
ODCH	ODMA Dedicated Transport Channel
ODMA	Opportunity Driven Multiple Access
OFDMA	Orthogonal Frequency Division Multiple Access
ORACH	ODMA Random Access CHannel
OSI	Open System Interconnection
PC	Personal Computer
PCCH	Paging Control CHannel
PCH	Paging CHannel
PCM	Pulse Code Modulation
PCPCH	Physical Common Packet Channel
PDC	Personal Digital Cellular
PDP	Packet Data Protocol
PDSCH	Physical Downlink Shared Channel
PDU	Protocol Data Unit
PICH	Paging Indication CHannel
PPs	Partnership Projects
PRACH	Physical Random Access Channel
PS	Packet Service
PSCH	Physical Synchronisation Channel
PSTN	Public Switched Telephone Network
PUSCH	Physical Uplink Shared Channel
QoS	Quality of Service
QPSK	Quadrature Phase Shift Keying
RA	Routing Area

RAB	Radio Access Bearer
RB	Radio Bearer
RACH	Random Access CHannel
RAN	Radio Access Network
RANAP	Radio Access Network Application Part
RAU	Routing Area Update
RF	Radio Frequency
RLC	Radio Link Control
RNC	Radio Network Controller
RNS	Radio Network Subsystem
RNSAP	Radio Network System Application Part
RRC	Radio Resource Control
RRM	Radio Resource Management
RSVP	Resource reSerVation Protocol
SAN	Satellite Access Node
SAPs	Service Access Points
SCCH	Synchronization Control CHannel
SCH	Synchronisation CHannel
SDMA	Space Division Multiple Access
SDU	Service Data Unit
SFIR	Spatial Filtering for Interference Reduction
SGSN	Serving GPRS Support Node
SGW	Signalling Gateway Function
SIM	Subscriber Identity Module
SMG	Spatial Multiplexing Gain
SMS	Short Message Service
SP	Special Purpose
S-PCN	Satellite-Personal Communications Network
SRNS	Serving Radio Network Subsystem
S-UMTS	Satellite UMTS
SW	SoftWare
TACS	Total Access Communication System
TC SES	Technical Committee Satellite Earth Station and Systems
TCP	Transmission Control Protocol
TD-CDMA	Time Division-Code Division Multiple Access
TDD	Time Division Duplexing
TDMA	Time Division Multiple Access

TFCI	Transport Format Combination Indicator
TI	Texas Instruments
TIA	Telecommunications Industry Association
TPC	Transmit Power Control
TR	Technical Report
TTA	Telecommunications Technology Association
TTC	Telecommunications Technology Committee
UDP	User Datagram Protocol
UE	User Equipment
UMSC	UMTS MSC
UMTS	Universal Mobile Telecommunications System
USCH	Uplink Shared CHannel
UTRAN	UMTS Terrestrial Radio Access Network
VAS	Value Added Service
VHE	Virtual Home Environment
VLR	Visitor Location Register
VPLMN	Visited Public Land Mobile Network
WAP	Wireless Application Protocol
WARC	World Administrative Radio Conference
W-ATM	Wireless-Asynchronous Transfer Mode
W-CDM	Wideband-Code Division Multiple Access
WRC	World Radio Conference

Index

Page numbers in italic, e.g. *207*, signify references to figures, while page numbers in bold, e.g. **163**, denote references to tables.

Index compiled by John Holmes